And I Saw A New Earth

2012 and Beyond

By Candace Caddick

Brightstone Publishing

First published 2012

Published by Brightstone Publishing
2 High Trees Road
Reigate, Surrey RH2 7EJ
United Kingdom

ISBN registered to me, the author, under Brightstone Publishing.

All rights reserved. No part of this work may be reproduced or stored in an information retrieval system (other than for purposes of review) or transmitted in any form or by whatever means without the express permission of the publisher in writing.

The right of Candace Caddick to be identified as the author of this work has been asserted by her in accordance with the Copyright, Designs and Patents Act 1988.

©Copyright 2012 Candace Caddick

British Library Cataloguing in Publication Data
A catalogue record for this book is available from the British Library

ISBN 978-0-9565009-2-2

Printed in the United Kingdom

With gratitude to the Earth

Books by Candace Caddick

Planet Earth Today: How the Earth and Humanity Developed Together and Where We're Going Next (April 2010)

The Downfall of Atlantis: A History of the Tragic Events Leading to Catastrophe (February 2011)

And I Saw a New Earth: 2012 and Beyond (May 2012)

Contents

Acknowledgements   vii
Introduction   ix

Part One: *Earth 26,000 Years Ago*
1. Bringing You Up to Date   3

Part Two: *Why 2012 Was Always Going to be Different*
2. Background Astrology   23
3. Earth's Preparations for 2012   29
4. Crystals on Earth   39
5. The Purpose of Avebury Stone Circle   42
6. Why Are We Here?   46
7. The New Earth   55
8. Moving From the Old Earth to the New   61
9. A Society Created by Lies   67
10. The Golden Wave of Energy   70

Part Three: *Life after 2012*
11. Get Ready to be Happy   81
12. Changing Your Lives   86
13. Life on a Living Planet   93
14. Global War on the Horizon   100

Part Four: *Days of Light to Come*
15. Earth Rejoins the Universe   111
16. One Soul, Seven Billion Bodies   120

17 Applying Light   125
18 A World of Slavery   129
19 Who Are the Light Workers Today?   135
20 An Army of Light Workers   138

Part Five: *Ascension*

21 The Story So Far   143
22 The Underlying Energy of Cities   148
23 Living in the New Towns   152
24 The Years Before Ascension   156
25 Ascension for All   162

The End   167

Books by Candace Caddick   174
About the Author   177

Acknowledgements

I would like to acknowledge the generous help I have received from Reiki Master Jean Jones for her copy-editing and commenting on the drafts. Without Jean's help I would not have had the outside voice every author needs. I am grateful to my daughter Heather who energetically checked the content of the book, and made sure that every word holds the intended golden light. The UK Reiki Association and my local Earth Healing group have continually supported me, as has Pippa Caddick and Justine Sharifian, and I am very grateful to them. I am also grateful for the support of my husband while writing this book.

Introduction

This is a channelled book about light, written by those who have ascended in wisdom and understanding and wish to help during a time of rapid change. We exist in timeless space, and see the changes ahead for humanity as it sets off on the path towards light and ascension. Currently many of you are afraid that your economies are in deep trouble and global warming or another natural disaster could bring your lives crashing down; you worry about your futures. We are here to guide and reassure you.

We live outside time, and tell you these are the single most exciting years you will ever have on Earth. You were born just to be alive right now and take part in them. Humanity is entering its golden years, when you begin to live as you always intended when you came to Earth. It will be like breathing for the first time, the sweet fresh air that is real life filled with joy, truth and clear-sightedness. Previously you have been hampered in your pursuit of self-knowledge and happiness. You feel that you have worked hard with few returns and lately it seems that you have to work harder for even less.

During 2012 the Earth receives wave after wave of light, enough light to change the way you relate to each other, enough light to show you the lies that have kept you from living in joy. By December the rebirth of Earth herself takes place filled with the energy of Spring and fresh beginnings. Humanity can use this energy to remove the institutions that failed to work, and restore the balance between work and play. 2012 is the end of the world you know: one of gross inequality and lack of hope. The years following give you the chance to build societies of love and fairness, and leave behind the institutions that failed you.

And I Saw a New Earth is written to be a guide to the coming decades, to reassure you that you can trust your intuition and your hearts, and that your real future lies ahead for you to enjoy. Humanity has one of the most important roles in the future of the universe.

A note on the subject of gender: so as not to be intrusive I am using He for God throughout. As far as I can tell this is the only planet that chose to divide into male and female.

Candace Caddick

Part One

Earth 26,000 Years Ago

Earth 26,000 Years Ago

1

Bringing You Up to Date

DECEMBER 2012; the month the Mayan calendar ends and the winter solstice aligns our planet with the centre of the galaxy. For the first time in 26,000 years the Sun rises in an unobstructed line to the light of the Central Sun. This periodic alignment with the centre of our galaxy has been a recurring part of the Earth's life cycle.

We are angels and beings of light and it is our wish to write a book about the changes to the Earth at the end of this cycle. These comprehensive changes will alter the planet you live on, and have the potential to change forever human society. When the planet you live on is reborn everything changes energetically. It is the end of the old and the beginning of the new world.

The last time your planet aligned with centre of the galaxy, which is known as the galactic centre, was 26,000 years ago. Life on your planet was focused in the oceans, and the seas teamed with whales, fish and other creatures. Whilst there was also life on the land, the more advanced communities were in the vast oceans that spread across a greater percentage of the surface. You live at the bottom of a "sea" of air, with birds flying through your skies, and you can imagine how it may have been to live underwater. Wave action has eroded all traces of these ancient cities, just as wind and rain erode your abandoned buildings. At all times they were aware they were given a home through the generosity of Earth and contributed to keeping their planet healthy. These underwater cities included an early form of humanity, where you experimented with life under the waves. If you had lived on the surface you would have been sidelined from the greatest variety of life forms, and you came here to interact and be involved. You are not alone in changing from sea to land and back again. Your time underwater brought you closer to the others alive here at that time, there were fewer barriers, and your homes did not contain windows shut to keep other species out. You felt connected to the Earth and others.

On the morning the Earth lined up with the Central Sun the water beings finished their lifetimes of living on this planet. They resigned their bodies, and by reforming back into their greater soul groups they moved on to their next experience somewhere else, but not as the same underwater species. They chose to try something new.

One species from the previous cycle finished at the ascended level. Those who reach ascension understand who they are and their relationship with their Creator. They have no need of further lifetimes to learn this and only return to Earth out of choice, not necessity. Humanity used the great changeover to move to a land-based body and new set of relationships with some new arrivals. None of the previous residents of the oceans moved onto dry land with humanity. You found yourselves starting a completely new game, but one without the natural ebb and flow of waves. The energy on dry land couldn't be more different, with the lack of movement and flow. Your feet were on the ground and when the wind blew it tended to come from one direction only.

This was a time of excitement, of changeover. Some of the species were staying on and they waited to see who the new arrivals would be, and how they had planned their learning experiences or games. You may have had that feeling at the beginning of a new school year when you were young. There was an adjustment period after the new Earth "day" started and the new species found their niches and learned to live with each other. Sometimes learning to live with each other means recognizing ones' position in the food chain. Being eaten is normal and part of the life and death cycle, and because animals have a shared consciousness they know they will be born again soon. There are lessons to be learned by eating others and being eaten, about the continuation of life and oneness with other animals. On this planet some species eat others, but rarely ever eat their own kind. When that happens it is usually because the species is unaware of their common soul. It's similar to eating your own arm.

The Earth had accepted a new group of souls, and the effect on her was refreshing and exciting. She made the adjustments in weather and geography that were necessary for these new residents and everything started again. Years went by and because of the nature of the human game she was overrun by dark angels, and tormented. She shut down in self defence and had a long, long nap, and was able to learn more about herself and the universe through dreamtime (as all living beings do). She awoke renewed in 2011 and spent one year gathering information about herself and her guests, and began to prepare for the new beginning in December 2012, and another new day. This time there will be a few major differences to do with what has been happening here over the past millennia, and the changes that are taking place in the outer universe. The most important is the wave of energy that arrived here at the beginning of 2012, which we'll talk about in Section Two.

The Earth exists to host games and all experiences are rendered back to God so that he can know more about himself and what it is to be alive. "God knows everything you do" small children may be told, and at the final end of the universe, he will. Everything in the universe has life because he is present in all things. He is the Source of life, and if he is present in you then you are a tiny (forgetful) part of God. When everything

Earth 26,000 Years Ago

returns to the Creator in the end, then each experience forms part of his knowledge of himself.

The light of the last Earth - Central Sun alignment provided energy, and a timer like an alarm clock. The Earth's day was starting anew, and there were species here that had contracted to stay until the new Earth day. When the time was up they moved on intact and whole, with many new experiences to share with the universe. There were other species, and humanity is one, who had different contracts and their time is up when a set of experiences is achieved and not when an alarm goes off. The human contract included working to their own ascended state, at which time they would join the Earth in her ascension. This is a contract that recognises that joining together is closer to God than staying separate. The timer part of the last alignment was mechanical, but the important, planet-changing event was the amount and quality of energy that pours through the centre of the galaxy to each planet when its turn comes to bathe in the light. 26,000 years ago your planet soaked up this light, and she was filled up to the brim with love and light, and shone with joy and bliss. Humans were living here at that time along with the other species, and received as much light as their bodies could hold. Sometimes there is a jump in the evolution of a species, and the amount of energy that arrives on these occasions is enough to push a species forward.

Back then, as now, the planet was coming to the end of the long cycle and was busy taking stock of her assets, arranging new games and contracts with new soul groups, and eagerly looking forward to a new day. Earth had already lived through a number of cycles and was familiar with the chaos of the end days due to transition. Her role in those times was to remain balanced and steady, and wait for the new day while making arrangements with new species. She has already hosted a number of life forms of which you have no knowledge that have unfamiliar shapes and colours, but these have all moved on and taken new forms elsewhere. Where are their dead bodies? They left no trace you can find now, but a memory of them remains in your Akashic records. These are the records of everything that has ever happened on this planet, species by species. It is a permanent library of the experiences of life that have happened here, and this information is available to be read by your Akashic readers. As the other soul groups share information already through their consciousnesses, you are the species that could learn the most from studying there. All planets have these Akashic records contained in the planet's energy fields. The universe is designed to share information, and what takes place on planets is not kept secret.

On the morning of the new alignment all the species waiting for that moment, stepped out of their physical forms and recombined into their individual soul groups. The energy of the Central Sun helped them to leave reunited and recharged by love. When a group is forced off a planet at the wrong time, let's say by being hunted to extinction, they often leave in fear and confusion. Their experience has been interrupted, they have a different outcome and it affects the development of their soul group. The next experience they planned to have on another planet is now altered to reflect their own changes. As beings of light we value souls being given the chance to be united in love, and these souls have been robbed of the opportunity to learn more about themselves. Hunting a species to extinction changes the web of life that exists here; it makes a hole and it is harder to balance all life (as you know from your earthly studies of ecosystems.) When a soul group is driven off prematurely, it affects all remaining life. Every species must then try to live in a balanced way on a planet with a piece of the ecosystem missing. If a species stops living in a balanced way their game swings more and more wildly, like a top wobbling before it falls over. It is easiest to see this instability in your own species where imbalances are threatening to topple your institutions. A fresh start will contribute to a more stable base, and from stability you can look to regain balance for yourselves.

As the planet's sunrise of 2012 approaches, you are still here and ready to experience it again, but this time it will be a little different. Periodically during the course of the existence of this universe there has been direct contact with the Creator. At certain times He has looked into the universe and taken stock of what He sees there, learning about Himself through all of our experiences. On 11.11.11 a scheduled inspection of the universe was made, and this time it was Earth's turn to have an unobstructed connection, a direct link from you to the Creator through the centre. This visit was planned so long ago that your method of numbering years was arranged to arrive at the 11.11.11 figure in order to enable the energy present in numbers to help you. Every time you see a clock at 11:11, remember the existence of spirit. A hundred years ago at the end of the First World War many countries adopted a minute's silence on 11.11. at 11:00 a.m. There was therefore a momentary space for silence when the true 11.11.11 arrived. Many here connected with their Creator and came to a new understanding of their relationship with Him. It mattered a great deal to some people, and not to others, a pattern that will continue in the coming years.

(For those of us who felt visited by God, what did we experience? For me, Candace, it was the vastness of a being that was surrounded by universes each the relative size of a small paper lunch bag. I also experienced an exchange where we swapped places momentarily and I looked back into our dark bag filled with stars through his eyes. And why not? If everything is God then I am also part of Him.)

It is not important that some benefited and received new understanding on that day, and some did not. As long as there

Earth 26,000 Years Ago

were some of you who allowed yourselves to be touched by God you became able to change others. How? You no longer have the same energy as before, and you walk through your days changing everyone you come into contact with. This is about the single human soul, and some parts of that soul have now changed vibration.

You are entering a new phase where the rate of vibration will be the most important part of your existence. The planet you live on is changing her vibration, and you are racing to catch up. She is a single unit, and you are working with seven billion people with their souls at different stages of development and vibration. Earth has a program of her own and she is carefully changing herself in an effort to take everyone along with her. Some of you have realised this and are changing as fast as you can to match her step by step. These are light workers, showing everyone else how to shine with light and love, and how to find the vibration that Earth is moving into. The light workers will anchor the rest of humanity as it catches up, and while this may begin slowly it will finish at a run. Who are these light workers who stride out into unknown territory where no one has gone before? You are one if you are reading this book, as are healers of all descriptions, and also those who love to laugh and spread joy to others.

The coming years will be a time of consolidation, of many beings living at a different rate of vibration. There will be more urgency to catch up across all the species, and to bring their vibratory levels in line with the Earth. As this happens there will be a buzz of expectation, of wondering what will happen next, and when it will take place. The aim of all the species on Earth at this time is to learn to live together, to combine their lives into a successful game. Animals have learned a lot about each other, about self-preservation, and about living as part of a greater whole. You watch them and wonder about their different ways, but you project your own feelings onto them. Your attachment to this life is far greater, as you see no future life and no further experiences. They see an endless round of lives and learning. Some animal souls are sad when they see their numbers die out, but they know it is not the complete end of their experiences. One day they will try something new somewhere else. The rest of life here knows that this cycle on the Earth has the potential to be very special.

You are embarking on years of differentiation, where some of you walk steadily uphill on an easy path towards the light, while others plod along below never realising that everything has changed. People will continue to be born and die, and one day in the future there will only be people walking gently on an uphill path and the road below will no longer exist. Every life has value no matter how it is lived (important!), but the need to plod on below will have vanished. Times change, and pathways alter, and what worked for one generation will not work for another. This temporary split begins now as the energy changes on the planet. You will live side by side with people living in the old world while you walk gently into the new.

You may look around and say to yourself: "Things are terrible! Young people are hopeless! What's the world coming to?" First, you have seven billion people living on this planet right now – why? This has a lot to do with impact; the things you could get away with in a small population aren't possible with a large population. When there are only a few people a little polluted water is not a big a problem, but a world overcrowded with people and no water to drink is an enormous problem. This makes a point that no one can miss and hastens the search for balance. (What a very interesting way, humanity, to go about solving a problem!) Second, you have all your best people here right now. From the oldest to the youngest, every soul incarnate here was ushered to the front of the line while others

Earth 26,000 Years Ago

stood back. If you see someone you feel you have nothing in common with, you both have this in common: you each bring something crucial to the game being played here on Earth.

When someone commits a crime other acts are set in motion like ripples in a pond. As unpleasant as some of those ripples may be, the criminal is playing a role that advances all of humanity towards the light. When you look in his/her eyes you are seeing part of your own human soul looking back at you, and you are looking at a person who may be doing what they came here to accomplish. You all have a chance to plan your lives before you are born, and we can see what wise choices you have made! Your lives weave around each other in the most astonishing way, like a complicated dance. Because your lives are forming patterns they make a whole in a way that might surprise you. It doesn't mean you have to alter the events around the criminal, because they may have come here to experience prison or being a victim. But it does mean that you are looking at a part of yourself incarnate in another body. A single overarching human soul learns by all the experiences that each member encounters. Your kindest act to yourself is to love and respect them regardless of how you are accustomed to feel about them. Your own personal growth depends on breaking down the barriers you have erected between yourself and other human beings. You will seek out experiences from now on that will teach you the most about yourself, and the light on the planet will be strong enough to show you all of your dark corners. Whatever you need to work on for self-knowledge will become clear.

Planet Earth, and each species present here, experienced contact with God at the same time you did on 11.11.11. There was a feeling of "its Earth's turn now" similar to the beginning of a race. Earth was waiting for the signal to start and gather herself for the next stage.

Earth 26,000 Years Ago

On 11.11.11 small adjustments were made, similar to the attunements received in a Reiki class where the student's native ability to receive universal life force energy is enhanced and stabilised. The attunements Earth and all life on her received that day prepared each being to enter 2012 and fully utilise the year's new energy. This is the great year where the golden wave of energy travelling across the universe washes over the Earth with the strength of a fire hose. She is braced to receive it, and will be cleaned of dark entities, dark residual energy, and anything else the entities established here for their own purposes.

In the long history of this planet there have been five previous soul groups that lived here and learned everything they needed to know to comprehend their relationship to life in the universe, and the One who created it. This is a very good track record for a planet, and shows that she is at the top of her game and knows how to host other species. But it is not all about the life forms contracted to live here; she has her own path to travel and her own needs. She learns alongside the life that is present on her surface, and she is affected by everything that takes place here. The energy of the actions, whether it is angry shouting or blowing a hole in the ground with explosives, radiates energy out to the surrounding area. She has to live with a blanket of this unpleasant type of energy on her surface like you might have to live with a thick coating of mud that you were unable to clean off. Every time you washed some away, more appeared. This type of energy tends to hang around, and one of the beneficial outcomes of healing groups is the energy changes locally wherever they take place. The light drives away the darkness exactly the same way as turning on a light in a darkened room, and for a time the area is clearer and there is less support for dark actions.

Healing groups were considered a vital part of past societies to lower the amount of dark sludge that accumulates on the

Earth 26,000 Years Ago

planet. Atlantean Reiki healing circles took place weekly where the entire village joined in. They took place over millennia and were a way of supporting the Earth with light in return for her help, and the people benefited from taking part and channelling the Reiki through themselves. It was only when these fell into disuse that Atlantis failed utterly. (We use the word Reiki here, but it would have had an Atlantean name then.) Any kind of healing can be used in Earth healing circles: Reiki, Spiritual Healing, Shamanic, chanting, etc. and as long as the intention is light it will help clear the area. This is often challenged by those who either can't see the higher dimensions, or want to control access to Spirit. The steeples on churches are designed to broadcast whatever is taking place inside, and for this reason it is important to feel what is happening in there. For thousands of years it was fear and control, and has only recently changed to talk about love. Music is a vibration that changes everything it touches. Live music made by singing or playing instruments is extremely powerful, but the music in churches is muffled. Its greatest effect is to make a clear space within the church walls.

Music rises out of the Earth itself and varies from location to location. To listen to the sound of the Surrey Hills in England you need only to listen to the music of Ralph Vaughn Williams. The American deserts have a waltz beat found in traditional cowboy music, the Alps sound like beer hall songs. The drumming that you hear around the world is played by those who may not have sophisticated instruments, but they can hear the sound of the Earth. Russian music has a very distinctive sound, as does Chinese. Composers may hear the local song of the planet on a higher level and not realise it is the origin of their inspiration. Sometimes it's easier to hear through the trees and bushes growing in the area, it is a part of their overall energy similar to a local dialect.

Earth 26,000 Years Ago

If you walked around wrapped head to toe in a thick blanket you would have trouble connecting to those around you. The layer of dark energy around the Earth has prevented her from being in contact with other planets and stars. This year she will be cleared of this energy and by the time she aligns with the Central Sun of the galaxy this blanket will have burned off. Life on her surface will be exposed to the same waves of light and be burned clear of old, residual energy that lingers around bodies and homes. This will be most beneficial to humanity, because you have accumulated darkness around you without noticing it was there. Very few animals, insect and plants have this problem as they see energy clearly, and it shows how much they love you that they will live in your houses and help change the energy inside.

The end of 2012 has the potential to be a fresh start for everyone, with an opportunity to let go of the old and familiar, and to trust that letting go will lead to something better. If everyone is hanging on to everything as hard as they can, all of this energy will not be released. You have a saying that goes "better the devil you know." This belief is how you end up with the "devil", instead of moving far away from him and letting him go. In the months of 2012 you have the opportunity to practice letting go and seeing that it is safe. If you don't hold on to fearful and painful energy it will cease to be part of your life, it will have lost its anchor. This is your main preparation for the events of December, 2012.

The Earth is our beloved planet, and we angels learn alongside her as she learns. We observe, and we feel what she feels. This planet has remained in her physical shape for a very long time now, longer than she originally planned, and she is ready to move. She needs to have your game on her surface concluded before she can stretch and reform. You agreed in your contract that if you can reach an ascended state in your

Earth 26,000 Years Ago

game you will join with other species on Earth, and ascend with her as one. Ascension as a united group of souls will put you all in a position to learn more about yourselves and each other than you could possibly learn separated as you are now. Your many lives' experiences would be shared between you, and your perspective and knowledge would be immense as you blend together. It will be a little like a computer game where you are able to move up a level and continue playing.

This is a unique joint project in the universe, and a very complicated game. Your human soul will assist another soul group to ascend, and all the pieces were provided in the beginning to help it happen. Every life form has a role to play to help you all achieve this, which is why the extinction of a plant or animal makes it more difficult. As it becomes more difficult for you, it becomes more difficult for the planet. On a planet crowded with species can you visualise the weaving together of all these lives to make a single whole dependent on each other? You used to be aware of this, and one of the last places it was remembered is in the Native American traditions of respect for Mother Earth and the animals they hunted.

Life consists of experiences to lead you to understand and know the one who created you. Your game was designed here within parameters to take humanity to this place of understanding, and the Earth has reached this point and is waiting for you to join her. She will be able to shed the physical body she is wearing and choose a new way of expressing herself. She will exist as pure energy in the future. Like all beings she is an energy body clothed in a physical form chosen to carry out her current role. She knows she could dissolve, but she will wait until every species is ready to move with her. You all have the same goal, to reach a stage of ascension and full knowledge of yourselves right across your species. At that time you will all be ready to exist as souls of light.

Earth 26,000 Years Ago

The Earth aligning with the Central Sun is like you waking up and showering every morning. You clean yourself with water, and she cleans herself with light and recharges her batteries at the same time. This cleansing takes place during the early months of 2012, continuing on through December 21st when the inflow of energy will begin to decline. It doesn't happen all in on one day in December! She survives on light and is assisted by your local star in a partnership, but this is her unobstructed connection to the central source of light, the light that also feeds into the Sun. In March 2012 Earth experienced the greatest solar flares from our Sun ever witnessed. These flares were for her, and had the effect of altering her frequency to a slightly higher level. People and other species had to step up their own vibrations to keep pace, and everyone seemed to have a head cold or other temporary illness. It was the first step of many to come during the year. The largest wave of energy arrived the third week in April to sweep forward those who were prepared to move ahead in their lives.

Some humans will reach ascension first and help the others to ascend together in a group, and Earth will not be able to join you until you are all ready. Some do not want this to happen. There are two separate teams of angels working in this universe and one of them prevents souls ascending whenever they can. Dark angels pull ahead in their game if they can get souls to descend into misery. Those games usually end in darkness and stagnation. Meanwhile the human game on this planet has demonstrated the full range of light and dark in a dramatic and sensational way, showing how far they can be pushed apart and still keep playing as a single game. It has been exciting and unpredictable to observe.

Washed free of darkness Earth will be strong again and reinvigorated, and able to breathe freely of the energy she needs to thrive. Her light will spill out and be visible across the universe once more. Think of vibrations of light being the same as singing, and her voice was silenced as she was slowly being smothered. Earth will be welcomed back to the family of planets after a long, long time, and that will make the planetary family stronger by being more complete. There is a different energy in any group that is whole and complete, and it is weakened by missing members. They have missed her strength and ingenuity, and look forward to having her back. When planets are strong they are heavyweight members of the partnership of light in the universe. Everything makes a difference in the overall game.

Now think again about turning on a light and driving away the darkness in a room. The newly-cleaned planet becomes the light, and the miasmas of darkness burn away. The years after 2012 will carry the energy of freshness and flow, and this will affect everything alive. You are approaching the first quarter of the next 26,000 year cycle, and it will energetically resemble the natural year that begins with the new growth of spring. The first half of the next 26,000 years will be like spring and summer before the change to autumn and winter. All seasons are essential for life.

In spring the plants pop up out of the ground, and burst into leaf. Earth is experiencing the quiet before the start of vigorous springtime energy where everything here will happen with greater and greater speed. It may take a few years for changes to settle down, and some of your man-made institutions will have been discarded. These years will provide energy for leaps in getting to know and understand yourselves.

Why does Earth need help from the Central Sun at all? What is the relationship between the planet and those who live on her? How did she get covered with tarry darkness? There was a time before the beginning of all the universes where the Creator existed, not quite alone but as a member of a collective of beings.

Earth 26,000 Years Ago

He rested in timeless space, and just lived and contemplated what it was to exist. He took the decision to explore himself by creating a great number of universes in the space around him, and filled them with small particles of himself and established the ground rules. One of the rules for this particular universe was polarity, an exploration of existence through the opposites of good and evil. There is a universe-wide battle raging from the very beginning between the dark angels, whose role is to create misery, and the angels of light who strive to create joy. By allowing freedom of movement and choice and not directing every action the Creator learns about himself. He has created waves of universes over and over again, and in timeless space there is no limit to how many times this can happen, or how long each round will take.

The Earth was contented in her role of host planet, until the human game started to falter. Humanity was the first race to live here who chose to be blindfolded to the higher dimensions and this left them vulnerable. Dark angels found they were able to settle on people and not be seen and live off the energy they provide, and control them. You can identify these people by their actions, not by what they say. The more they could create the energy of fear and unhappiness, the more they thrived. This is how a coating of tar enveloped the Earth; a dense, non-moving layer of fear, a scum of stagnation and sadness. They eventually felt quite at home here (plenty of misery to feed on!) and ran the planet through human beings to suit themselves. Very few people over the years could see them to oppose them. The influence of demons on humans was known through many societies around the planet, but in the scientific Western cultures there was no measurable proof of their existence. We would like to point out that science is inadequate to measure good and evil, and they do not take it into account. Science needs people with good and true hearts in charge of research.

Earth 26,000 Years Ago

We angels and beings of light never abandoned you or this planet, but at times we had difficulty reaching her. We were able to help her through influencing people in our own way, through instruction and supporting those who worked through love. We knew she would have some help from the universe in 2012 which would also assist you. This is the time she has been waiting for, the passing of the old days and the beginning of her New Year.

Part Two

Why 2012 Was Always Going to be Different

2

Background Astrology

IN THE FIRST section we spoke about an event that can be calculated and is part of the visible universe. The Milky Way rotates and once every 26,000 years returns to the same place in its orbit in relationship to certain fixed stars and galaxies further away. In December 2012 your Solar system will have an unobstructed view of the Central Sun of the galaxy. If you think of a large ball of stars moving in an carefully choreographed, intricate dance around a central star; once every 26,000 years each star has an unobstructed line straight to the central sun. Each star and solar system has some time when they receive the light directly before moving behind another star that then has their opportunity. The Milky Way is flat but the principle is the same, each star and solar system has their turn in the light. In December 2012 the Sun will appear from Earth to cross the line of the Milky Way, which will be difficult to see in the daylight. It will form part of a cosmic cross as it appears to move from one side to the other.

The Universe is a whole and complete system, and it is the interrelationship of planets to their stars, stars to their galaxy, and galaxies to each other that make up one universe. You are used to looking at 1) yourself, 2) your planet, 3) your solar system, 4) your galaxy. If you focused that much on your little finger and failed to see it as a part of a complete person you would be missing a very large point. We spoke about the way each human life weaves around each other and makes up one whole experience for the human soul. In the same way every galaxy (there are so many of them!) dances around each other and makes one universe. The universe is made up of life and experiences, and one day humanity will share with others what they got up to here on Earth. If you sold tickets to that talk you would be overwhelmed! Your experiences would become part of the background knowledge of every consciousness, and taken into consideration the next time any species planned a game on a planet. "Humans already did that separation/blindfold game, let's do something slightly different." Humanity pushed the boundaries in a certain direction in this game, and no one is going to have to repeat what happened here. What happens on planets are the building blocks of experiences. The games on other planets affect you even thought they seem so far away because everything is interconnected.

When the Mayans went outside and looked up at the night sky it was filled with stars from the horizon to those high overhead and the Milky Way looked like a spilt ribbon of milk crowded with bright stars. The night sky was filled with planets and stars, just as the daytime sky was filled with birds, trees and leaves. We would have to go very far away from towns and cities to begin to see night skies like this today. Astrology is about what people in the past could see when they looked up, and about what they observed in their natural world and how it related to the stars they saw. We're reminded of a greater whole, and that Earth exists within a larger universe of friendly stars and planets.

What will you experience in December 2012? You will experience the light behind the light of our Sun. A very long time ago some stars in each galaxy took on the role of light keepers. These are the biggest and strongest stars in the centre and they hold the most light, which in angelic terms means they hold the most love. They are able to create and maintain an environment of light for other stars and planets to live in, and they keep darkness at bay. If they weren't there anchoring light, the universe would be a different place as they support smaller stars with their light, like relay stations on Earth for electricity and telephone transmissions. There is no limit to light, as there is no limit to love. It is this energy that Earth will be exposed to in 2012.

People on Earth have a viewpoint of the Central Sun that no one else will share, because it is your turn to be in alignment with the centre. Going back to scientific astronomy you will realise that these events are not momentary alignments at the dawn of one day, but movement that takes place slowly over long periods of time. You began to conjunct these particular stars decades ago, and there are societies on Earth that started to feel changes over the past seventy years, as there were no obstacles between them and the stars, physically or culturally. Usually these societies are considered by you to be "primitive." They don't have layers of civilisation dulling their senses and the incoming energy was able to work with them. You will all have your opportunity to catch up.

Now we are going to talk about a larger universe where stars and planets live their lives according to their own choices, and each has an individual name or signature energy. All those stars and planets out there feel different and send out a different vibration. That's how we know them, by their native vibration. Each planet has something different to offer because they are not the same, just as children born in a family are not the same. The closest body in space that affects the way humanity feels is your Moon. Emotions are heightened during the full Moon, and the police know they have to keep an eye out for trouble. You cannot exist in isolation, or fail to be affected by all the planets and stars in your environment, and they in turn are affected by you. All vibrations ripple out and are returned back to you as a reflection of your lives now blended with all of life. It is how the universe was designed to work, and that's why your game of separation from each other and everything else is so extraordinary.

Why 2012 Was Always Going to be Different

You are attempting to live in complete isolation from others and even from your greater self.

When stars and planets group together they can work to provide a larger energy. If you have a trio of singers, they will be quite different from a large choir. A male choir will make a different sound from a mixed choir. Sound is energy, and the grouping of stars is similar to these different choirs. The cosmic cross is a specific grouping including the bright Milky Way and your Sun. Your local choir, by the way, is your local solar system. You are joining the biggest choir in the galaxy as guest singers. Being immersed in sound, as long as it's not sharp or flat, is good for healing and balancing, and smoothing out all the rough edges. The Earth is looking forward to participating in the cosmic cross choir, and it will focus energy on her. Being surrounded by this energy has an incredible effect on your entire solar system. Hum to yourself, a nice vibrating rumbling hum. Now think of the Earth humming outwards and the galaxy humming with her; this is not something that is done to the Earth, it's something she takes part in.

This is a very Earth-centric book, but we want you to know that the planets in your solar system each have their turn in this energy also. Your part of space is busy with souls coming and going from different planets as the changeover day arrives, just as on Earth. If you had not chosen to be blind to the higher dimensions you would see all of this activity. It's a high-energy time, and the energy coming through is exactly what is needed to make these changes flow with ease. It's routine for all the planets here, but there is a buzz of excitement over the coming cycle, a feeling that the Universe is changing direction and heading for the end.

The purpose of this year's conjunction is for the stars and planets of the galaxy to support the Earth in the same way you connect into your friend and family groups and support each another. There is an interlinking web of many connections, and these heavenly bodies understand their mutual relationships. If one planet or star is suffering they are all aware of it and it lowers their well-being. They remember that they are one soul group and that caring for each other is the same as caring for themselves. Helping their weakest members raises the vitality of all the planets.

Earth-based astrologers focus on your local planets and Sun, and study the effect of their vibrations on people living here. These planets in turn affect all life here and the planet itself, and the Earth in turn affects them. The greatest influence on the planets in your solar system is your Sun followed by the stars. Although some of you are aware of this you have little knowledge of the individual energies of the stars and how they affect the planets and yourselves. If you feel drawn to sit and meditate on your Sun (don't look straight at it, sit in a patch of sunlight instead) you may come to learn more about them; there are so many stars making up the universe. Stars are reservoirs of light in each galaxy and bring that vibration into their surrounding areas. They are not small lights in the sky, but enormous, powerful beings each surrounded by smaller stars and planets that rely on them to light up the darkness. This they do, each protecting their own space and working with other beings of light.

Astrologers interpret the vibrations arriving on Earth, and allow people to consciously take advantage of what is happening in their lives. Wisely many of you are aware that you have a Sun Sign, and may know nothing else about your astrological birth charts. The Sun is your strongest influence, followed by the Moon, and it is the giver of life on all planets in the solar system. Without its light, warmth and energy you would be cold and lifeless. The energy of this particular star has imprinted these planets and all life here, and if you travelled your origin would be known anywhere in the

Universe.

There is another reason why any Earth-born resident would be known across the universe, and that is that we are familiar with Earth. She is unique, and you are not only living on her surface, you are practically small extensions of her. She has her own name and feel, and you have absorbed this from her, she gives you life. The planet whose vibration has the greatest affect on humanity is Earth. Earth's children is a good name for you all, and is something you share in common with everything here. If a bird was alive on another planet it wouldn't be the same. When we say "be one with everything", you are living with fauna and flora that you are almost one with already, and you need take only a couple of more tiny steps to realise this, and enjoy the relationships. Different forms of life on this planet share a great deal of DNA, possibly as much as 90% and higher. It's not an accident that this is so.

The vibrations penetrating the Earth from the Sun and other planets are attuned to her more than to human beings. They come from her people, and they resonate with her and she understands them without interpretation, the same as you do when speaking your native tongue. You are trying to understand the language of the planets in the same way as learning a foreign language, and some people are very good at it. But so often in the past the vibrations have arrived for her and been blocked by an impenetrable layer of darkness on her surface. She has been isolated and unhappy. Your various electronic transmissions of TV and radio, and phone masts don't particularly bother her; they are puny compared to what is arriving for her from outside. They have a far greater affect on people by distorting the clean energy of the natural world.

3

Earth's Preparations for 2012

THE VIBRATIONS from space enter the Earth from all directions, and she normally contacts them just by existing and vibrating her experiences outwards. Connecting through the great stone circle at Avebury she can amplify and focus her outward communication. It's been a long time since the stone circle at Avebury was in working order, and a planet covered in tar is silent. The only reason the Earth did not suffocate was that she was able to use a few natural areas of protected beauty as breathing spaces. In *The Downfall of Atlantis* the Lake District in northern England was mentioned as the largest of these, kept open by the love of people for their land. The actions of those living there protecting that area kept it unblocked and free. Natural places full of life, some no larger than freckles, were kept free of darkness by local people. She can feel the love people have for her in those places.

You have chosen to live blindfolded to the higher dimensions, and it is part of your experience of living here as a soul group. We find your game so difficult that we cannot imagine how you are able to live so alone and isolated from each other. In spite of that you are capable of such acts of love and connection that you give us great joy and hope. All of you who act out of love have an enormous impact on your world, we can see the light that shines out when that happens and we admire you for it. You are brave and wonderful and right now you have the care of a beautiful planet in your hands. This is the point you have reached after many, many millennia. Your actions are being watched by the universe because you can choose your next direction, and it is not possible to predict the outcome. You continue to develop and have reached a very interesting point in your long-term game. If enough of you choose a direction, you will begin to drag the others behind you in your momentum. Which way do you wish to go?

A long time ago you settled in an archipelago and we helped you form the many civilisations of Atlantis. We originally thought we would help humanity by hot-housing those living on the islands, teaching them and sending them out into the world later. As teachers the Atlanteans would instruct and guide people who were living in the less advanced societies on the mainland. Thus there would be a group of enlightened teachers leading the way and all the other peoples would follow them, and the game would end as humanity ascended to light. That was our plan and we thought it started well, and then Atlantis descended into darkness instead. The civilisation had to be removed as an emergency measure to protect the Earth. (We wrote this sad story in *The Downfall of Atlantis* so that you would remember and not repeat the same mistakes.) This all happened so long ago, and it has taken a long time to reach a point where everyone is ready to act, as human society has gone as far as it can in the old direction. Like a pendulum on a clock you have swung as far as you can in one direction and now there will be rebalancing and change. Not overnight, but this irresistible process is beginning now. Exciting times are ahead for all of you.

We find it interesting that you have reached a point of reversal and change of direction in the last few years just in time to take advantage of the new energy. One of the biggest events has been called the "Arab Spring" in the west, and it seemed to begin so suddenly and then escalate to topple governments. We look at currents of energy and see the force of the universal wind that began to blow. If you had a dilapidated shed and a very strong wind blew it over you would not be surprised. These regimes were in the same condition as that shed. The energetic wind began to blow in 2008, and has been steadily picking up strength. The financial crisis beginning that year resulted in the rotten structures of the banks being propped up with unimaginable amounts of money; you now have rickety banks propped up with banknotes. That is not a stable structure, and the universal breezes are getting stronger. Do you think they will stand up to hurricane force winds? Anything that is rotten and is blown over does not deserve to be mourned, and it will create a space for new and healthy growth. The universal winds that blow are golden with light and can be trusted to support healthy growth while tearing down decaying institutions. This is change with a helping hand from the light of the Universe, and not a time for worry or fear.

There are a few pieces of advice for the year 2012: let go of everything that no longer serves you, and hang on tight as this rollercoaster ride really takes off. Each year has a consciousness and an energy that is unique for twelve months until it fades away on New Year's Eve. These years are coloured by the energy present in numbers that are assigned to them, and we are only talking about years starting on January 1st in this book. Of course, you can start counting a year from any day you choose. 2012 is the year when the winds of change blow, and it has a joyful, confident personality. It's going to have the same effect as a friend running up behind you, catching you and making you all run forward together. You will run or fall over, and you will be able to run further and more quickly the less you carry in your arms. This is a time to remember that the universe will support you if you let it, you don't have to ask but you need to stop blocking it. The new Earth is worth running to get to as fast as you can.

Why 2012 Was Always Going to be Different

Reaching the new Earth will be like a windy day where you walk and pick up speed as the wind pushes from behind. Sometimes without even noticing it your feet start to go faster and faster with the force of the wind. As it becomes stronger you start to run in front of it, laughing. You see someone fall over and other people hanging on to lampposts but you keep running ahead. Later as the wind dies down you find you are walking safely with others in a soft and beautiful world and you got there by doing what came naturally, running while being pushed from behind. There are many out there who will begin to run and not understand why, but they will find themselves walking with you in a new world of incredible beauty and light. The universal wind will have blown away the cobwebs and veils that hide this world from you.

This is a book about living on that new world, that soft and loving world that is the new Earth. Whenever you see a part of the world that takes your breath away with its beauty and health, the vibrancy of the plant growth and the freshness of the breeze, do you wish you could see an entire planet like that? A planet healed of all its hurts and scars, with the irregular wobble of the seasons predictable once more? The Earth restored to the planet she was and will be again? This year is about resetting this planet back to the beginning of the day when she is in charge of her own body. There is excitement and relief from the other species at the prospect, and they don't realise that you are oblivious to it all. Some of you will arrive first, but in the end all of humanity will understand what has taken place.

The year 2012 shares the cosmic cross in common with the events of 26,000 years ago. The Milky Way has rotated once again, and we could go back and back into the earlier rotations and talk about events then, but we think you get the general idea, and we want to move forward and not back. Your galaxy exists in a universe of spinning galaxies, and each time it rotates to the same point the outer universe has altered some of its positions. Not by a lot necessarily, but they have changed. Go back to the idea of a paper bag holding this entire universe, your galaxy does not spin in one place, it has been circling around the inside of the bag. This is far more important than a single rotation of your Sun inside the Milky Way, as the background influences on the galaxy have changed.

As your galaxy circled the centre of the universe it has travelled through the universe's astrological houses. You use similar houses on Earth in astrology. The universal astrological houses each contain a unique energy that overlays everything else that is happening. It acknowledges the vibrations of the stars present in the houses and how they change the energy that surrounds them. Moving into a new house presents a new set of opportunities to learn about yourself from a different angle. We have a system in place to organise the universe and this is one part of it. You are leaving the house of winter, sleep, dreams and internal focus and are moving into the house of spring, softness, the virile strength of youth and new life. Earth has been dreaming for over twenty thousand years.

The universal houses teach everyone and everything the lessons that are central to that house. The Earth has been sleeping and dreaming, and her long rest has taught her much during her altered dream state. You also sleep and dream, and that is when your mind relaxes and you meet us and learn from us. Your mind has to let go of daytime concerns when it gives way to sleep. We bring gifts of learning when you are asleep, and some of you do more work at night than you ever do in a day, you are your true selves and not limited to a physical body on one planet. You travel the universe in our company and sometimes do brave deeds. Then the dawn comes, you wake up and won't be our companions again until you sleep. Every now and then you meet one of our dark brethren and have a nightmare and it seems so real! It is real, and it is happening to your higher self while your physical self sleeps safely in bed. No wonder you wake up in relief.

How many of you are familiar with the way you learn while asleep? This is one of your key skills, learning something like a piano piece one day and your brain going over it at night. It applies to so many things you learn in your lives; you are very active once you are asleep. Each time you wake up you have learned something new and are not the person you were the day before. There is separation between the higher self who is busy learning at night and the physical person who puts it all into practice the next day. (That's a good reason for saying "I'll sleep on it" when needing to make a decision.) In a similar way you live on a planet that has just finished taking a nap for twenty thousand years. Her need for sleep was similar to yours, and her rest-time allowed her consciousness to focus solely on dreams of light and knowledge, and nightmares of despair. What does a planet dream of? She dreams of interactions with others and relationships, she dreams of freedom and adventure. She assimilates the lessons she learned while she was awake, and she is able to do this because she is not paying attention to all her waking duties, just the same as when you are tucked up in bed. The Earth is a lot like any living being, and she is travelling a path that many of you would find familiar, a pathway to self-knowledge.

Early in the spring of 2011 all the plants and animals were full of excitement at the waking of the Earth. Of all the life on the planet, only humanity had lost its ability to sense this. You used to know all about her when you were outside all day, but now that you live shut away from the Earth in cities and houses you have forgotten, and very few in the West now teach others this information. There is no reason why Western societies cannot learn to feel the pulse of the Earth again. Animals that grow thick fur coats before the cold winter and plants that have heavy autumn fruit know the weather in advance. One fine spring day the Earth woke up and lay quietly blinking her eyes before rising. This last year she has been gathering information about herself from those on her surface who talk to her; trees, plants, animals, elementals and more. A status update is a more modern way to think about it, but she wants to know about everything before acting. Think of it like this: when you sleep your higher self is out having an adventure but if you were to begin to roll out of bed your body would wake up, resume consciousness and take charge. The Earth's higher self has been out learning new things, and her very physical and solid body has been sleeping. To wake up, her consciousness had to return to her body and take charge of it again.

This is the big story of 2011, the return of the Earth's consciousness as she ended her long nap. The birds sang of little else, each animal and plant knew that she would wake and prepare to take charge again. We are talking about an Earth that none of you remember, an awake and sentient planet that runs her own life and is ready to move forward. It's a little like you throwing back the bedclothes on a sunny morning and leaping up ready to go; exactly that type of energy. She has recovered her joie de vivre, she's not tired or weak, and she is the mistress here. If you thought that you ran this planet she would say that you work together for a common end. She is ready to work and go forward. What steps would be useful for her to take now? She spent 2011 gathering information to work out the answers.

Now, almost a year later, she is ready to stretch gently in many directions and begin changing herself. If a vine grew along the ground branching in different directions and lifting a little dirt here, and pushing a leaf there it would reflect the rate and scale of change she is initiating. She knows where her points of balance are: how much rain is beneficial, sunlight, winter snow, etc. and she knows how to stretch out and reach those points. There may be some bumping and wiggling as she makes herself comfortable again. Earth has this year of 2012 to get to where she can begin to move forward confidently. This is the lovely thing about moving forward and not back, it doesn't matter what she is moving away from, it only matters where she is going next. The same applies in 2012 to you, too, and you will reach a point where you are comfortable in your own skin.

The Earth is a hybrid planet intertwined with the crystal consciousness, who was the first soul group to ascend from her, and the crystals remained behind to work and learn with the planet. She agreed to host the human game if we stayed here until our own ascension and joined together with her when we were both ready. This planet knows that the timing is right for joint ascension, and that the light created in this act will have far reaching effects. She has used the experience of hosting humanity as a time of learning her own lessons; you've helped her to understand patience, stamina, and flexibility among other things. She learned more from you than from anyone else because your own time here has not always been straightforward, and there have been false starts and problems along the way. You must be clear that learning from past mistakes is a legitimate and sometimes fast way to learn. These are extremely valuable lessons of the "I'm never doing that again" variety. When you are in the middle of a situation it can be hard to know what is really happening. Earth took some time out and had a nap, and has a fresh perspective on what is happening here.

On this planet right now you have humanity who is still playing its game of learning to find God even though it is wearing a blindfold. There are animals and plants that are learning about life, and exist in a web of co-dependence, with humans off to one side, connected, but not connected in the same fashion. There are many planetary elementals (gnomes, dryads, centaurs, etc.) that learn by serving the Earth and plants in the natural world. Animals and insects chose to have physical forms that move around, but they work with elementals and can see them clearly. Angels are here working for the light, and dark angels are here working for the dark, but unfortunately you almost never see either of us. The others who are here are very few and are observers, and they do not act. These many beings call the Earth home and live their day-to-day lives connected to her, listening to her voice and working alongside of her. There are many species that run away before the earthquakes or tsunamis hit and grow thick fur for winter. You have all the same abilities they do, but you seldom use yours. You only need to trust yourselves and your senses.

All of these beings have the potential to join Earth in her ascension. She's so big, why would she benefit from smaller beings ascending with her? They're not actually smaller. If you look at a person you can't see how big their soul is, unless you practice for a while. How big is the communal soul of all the ants on the planet? (It's larger than the physical size of Earth.) How big is the soul of the Earth? It's about the size of the ants' soul. Ascension is about souls, not about physical bodies, and each species is one soul expressed in many, many physical bodies. There is one human soul and you chose not to remember that to make it harder for yourselves, to make the game different from that of the other soul groups here. There are thousands of soul groups present here now, just to take part in this experience of combining and ascending, and learning more about themselves. The application rate to be here was high from the beginning, and the Earth selected those who could work together. No other planet has attempted to do this, and success was not guaranteed, however all experience in the universe is valuable.

The number of soul groups on this planet right now is staggering. You haven't travelled much and you may think that all planets are filled with life just like Earth. Most planets try to play host to games where the life forms are larger and there are fewer of them. Long before you ever came to Earth a blue bat-like being the size of a flying house once ascended from this planet. They owned nothing. You have a world of houses and rooms filled with little tiny objects; tiny animals and even smaller insects. Earth is boiling with life, and why would she take on so many soul groups at once? (You have to admire her organisational abilities!) There have been hints that aliens and others like to watch Earth; sometimes they do because it is literally the best show in the universe. We are watching her game of hosting multitudes of life: how complicated can she make it and still succeed? Where you choose to find God through separation from each other and unknowing, she is finding God through hosting games with more souls than you can imagine; we feel sure that this is her final game as a solid planet. Ascension will take her to a new playing field, united to the souls living with her. She will be one step closer to rejoining her Creator.

4

Crystals on Earth

EARTH'S FIRST resident soul group was the crystal consciousness. The crystals' time on the planet ended when they learned enough for their own ascension, after which they devised a way of working together with the Earth. The crystals are tied to her through love, and because of this their partnership forms a unique planet. You live on a planet with a dual soul in one sense, as the crystals live within the Earth and their thoughts are present with hers. They share together the same space and the same experiences while keeping separate identities. It's like a marriage where two people work to provide one home.

Earth was able to use crystals to heat Atlantean homes by recycling geothermal heat to the surface through their bodies. She provided all the energy Atlantis needed with crystals by using their powers of amplification and the purity of their structure as conduits. You use glass fibres now in superconductors, and they are of similar crystalline structure. Crystals from the Earth are alive, and man-made ones from laboratories do not have the same energy present. In the planet are many pockets of crystals, far more than you have any idea of, and part of their role is to communicate with the Earth. Many individual crystal voices are able to feed into the greater crystal consciousness, and provide information about life above ground in an instant. Information can also be collected and sent out to all crystals, and this is how some of you learn at night when you are asleep, by keeping crystals at your bedsides and wearing gemstones. They are small beings of love on the surface, and large building-sized ones deep underground.

Crystals played their part in caring for the Atlanteans until near the end, and then withdrew their help when threatened. This was one of the major twists and turns of your human game. Since then they have been busy underground and keep themselves informed while working in the background. Crystals have been going through their own period of quietness to go with the energy of the winter astrological house and are getting ready to re-exert themselves and help the Earth.

Earth began her working life in this universe as a gas ball hosting just one soul group, the crystal consciousness. The shiny and hard crystals you are familiar with were at that time gaseous, flexible and mobile. They didn't walk across the Earth, but as gas they moved through her. She had a lot to offer even then, and crystals learned who they were and their relationship to the larger universe, and to God. They became an ascended soul group and could leave this planet and move on. Out of all the souls that have lived and played here, they were the only ones to choose to stay with her. They went from gas to solid gradually as she cooled, scattered through her own form. They became fixed into place underground for the most part, and began a new period of assisting her in any way they could.

Crystals are preparing to be active again, more active than they have been for thousands of years. They are approaching the time of their re-emergence, when their withdrawal from events will end and they take up some of their old functions. Healers have been working steadily to unblock ley lines and repairing the mechanisms of the great stone circles and smaller local power centres all around the world. When the Earth is free of her dark coating the stone circle of Avebury will be able to work properly. It is circular so it can spin, and as it spins it pushes energy down the ley lines and the energy from the

Why 2012 Was Always Going to be Different

Universe and the centre of the planet will travel around the globe. Parts of Earth that have been inert will feel the flow of energy again and she will regain her former strength and control there.

Crystals have their roles to play in the coming years. You have caves deep in the Earth filled with giant crystals and these will connect with lines of energy. What was motionless will be active and alive, as energy has been absent and will be restored. This cohesion of all the worldly crystals makes for a huge presence of light shining up and out of the skin of the Earth. She's shining now, and when you add their light it will be very bright. Angels of darkness work in the shadows, through lies and deceit. When you are able to shine a light onto what they are saying they will have less influence, and it is important not to give them anywhere to hide but to uncover the lies and the dark places they hide in. The lack of light here from the Earth is one reason why they've been so successful. Just look at the endless warfare if you have any doubt about their success.

5

The Purpose of Avebury Stone Circle

LEY LINES ARE like acupuncture meridians for the Earth. The principle of health in acupuncture is that energy flows down the meridians in balance; not too much, not too little and no blockages. During the years when Avebury was neglected and disused, energy was barely travelling down her ley lines. This one circle has the major role in drawing energy in from the Universe to the Earth and sending it down the ley lines. The energy grid made up of these lines is part of a planet's structure; it is her energy body and is just as much a part of her as her physical body. Energy flows under the surface, and over the surface at the height of tree branches, and there is a final set quite a bit higher than that. The energy lines dive below the surface, circulate and re-emerge, and the energy flow is necessary for life. An acupuncture patient with no energy would be dead. The Earth woke up to find the first thing she needed to do was improve the trickle of flow. This is a good example of how healers help the planet, as the great and small stone circles will be ready on time to push the energy down the ley lines.

Avebury Stone Circle is a complicated, delicate and precise piece of machinery set into the Earth that collects a universe full of information. It is tuned into everything that takes place and is endlessly adjusting to receive fine nuances of information, as well as the broader events. The Earth would know if someone cut their toenails on another planet as the whole universe is made up of small individual events. At the same time she is sending information outward about everything that happens here. Earth receives energy like a broad funnel down through Avebury into the planet, and when she broadcasts it goes out in a straight line to be picked up by life out there. If you look below Avebury you can see a two way flow of energy. This is how the Universe remains one entity, by sharing all its information among its inhabitants. The Earth sends out information and is reflected back to herself by the universe, as we are reflected by other human beings.

Joined to Avebury is Silbury Hill, and they are connected like two gears. Silbury is the smaller cog, and as it rotates it turns Avebury and the machinery there. Below Silbury is simply the Earth, and she is the one turning the cog. Energy and information received into Avebury is spun out into the ley lines, and then into the rest of the planet.

The standing stones with all their different personalities are necessary because each one is an aspect of the outer universe, and they assist by pulling their specialist information through and feeding it into the centre. They look like hollow tubes. (I'm afraid this is all a gross simplification of what is taking place there.)

There is a better visualisation than machinery of what's happening at Avebury and that is the human face. When you are with people you are listening and using your eyes and ears to take in information from their faces, at the same time you are responding and giving back information through your expression. Avebury is like the Earth's face, and it is the only one she has.

The flow of information she is absorbing will change everything. It is part of our environment and we breathe it in, drink it in our water and eat it in food grown in her soil. We walk through a blizzard of information and some of it resonates with us, and teaches us.

Why 2012 Was Always Going to be Different

A laser is focused light and extremely powerful and versatile; it cuts through solid objects. This is reassuring, and should give you an idea of the power of light. Many people think that the end of 2012 will come and go, and nothing will really change. But once the energy changes, everything else will follow. Energy always leads the way, first the idea to act, then the action. Some events have more drag and resistance attached to them and will be slower to change; others are very ready to change right now. Your world will adjust item by item to start with, and change will continue as long as the energy no longer supports the old ways. The speed of change will pick up, but it will never be faster than is good for you or the planet.

Change is a wonderful energy, the most powerful in the universe. Resistance is a waste of time, like an ant trying to hold back a flood. Yet resistance to change is a deep-seated problem for humanity, and that leads back to trust. What has happened here to rob you of trust in the future? You could relax and allow the universe to give you shelter, food and all the good things that are there for you. Sometimes you look at a person and think they are lucky, but we can see how they allow good things to come to them. Is there a limit on love, and how many people you can hold in your heart? No. There is no limit to the number of good things that are waiting for you when you make space for them to come in. Make a space by believing that you are always loved and cared for. Let the changes occur by looking forward to everything becoming better. You are getting closer to finishing this level of play and moving up to a new, completely different set of experiences.

The change you are all waiting for, and rarely acknowledge to one another, is to do with hope. There must be a purpose for your lives here, you can't go through so much personal pain for no reason, and you have to have hope that there is a point to living. There is more to life than eating, sleeping and earning money, and the purpose of your experience here is to discover why you exist at all. You designed a game that would lead you to discover what its like to find out who you are from the perspective of an innocent baby. The point of your entire existence as incarnate beings is to learn that you are God, and this knowledge is achieved from a position of complete unknowing. When you understand that you are all a tiny part of your Creator, and that the universe exists to further His knowledge of Himself then you have finished your many lives here. After reaching that understanding, when you die you begin a new phase of assisting others to the same level of knowledge until the entire soul knows its purpose.

A long time ago your Creator was a being who was whole and intact and all-powerful. He felt the need to learn about himself by experiencing so many different things that He divided into pieces no larger than a grain of sand and filled up countless universes with Himself. Some universes lived to one set of rules, some to another set. Your universe is a universe of polarity where there is good and evil, black and white. This is the underlying basic condition for life here. When you look at yourself you do not see God, but that was what you wanted for your own set of experiences on Earth. You are here and exist because He gives you life. Rocks are here because a tiny part of Him exists inside each one. At the end of your time on Earth you will all realise that everything is God. And now, what percentage of the universe do you think may not know that they are a part of God? Perhaps less than one percent, much of which is made up of human beings. This is another reason why you are the favourite planet to watch, we keep waiting for each of you to go "Ah-hah!" and make the connection. We enjoy the suspense, we respect your journey.

6

Why Are We Here?

WE'VE TALKED about Earth and her part of the Milky Way moving from the house of winter to the house of spring. She is in the vanguard of this galactic movement; her sector is crossing over first to be followed eventually by the rest of the galaxy. You've seen stop-motion photography of snow melting and buds bursting forth from trees, where it looks like it is happening before your eyes. Once the thaw begins spring soon follows, and in this case we are referring to Earth's sector leading the way out of the deep freeze being just like warming your back at an open fire. The thaw will spread into the rest of the galaxy. Because the Milky Way is a single unit, any single part of it that changes has an effect on the rest of the stars and planets. In this case we are looking at the whole galaxy changing to keep up with the Earth and the springtime energy. The entire background energy of this galaxy will go from frozen winter to vibrant spring over the next one thousand years. Think of your response to spring every year.

This will have two separate effects on your daily lives on Earth. First she will not be experiencing the drag of the rest of the galaxy in winter as she goes forward with her sector, and everything here will happen fast. Second, that sleepy, hibernating energy will be gone (unless you are a hibernating animal). You have spent years contemplating yourselves in the winter dreamtime, and you are all ready to move forward. This is it; the home stretch is ahead of you, and its time to act instead of dream. The speed of forward motion will be astonishing to you, and you will look back and see that 2012 was a watershed year.

Taking action is where the thinking part of you meets the physical body and you become a single person existing in the present moment. Dreaming takes you out of yourself, and one of the reasons you are here on Earth is to be fully alive and living. Can you think of any other species that is not present in every moment of it life? Yet you are often worrying about the future or regretting the past. It's a habit. The winter energy that supports dreaming will vanish, and many of you will wake up and live in the present, and it's the only place you can live. Birds and beasts don't worry; they simply live from moment to moment. Some may assume that you are better than them because of all your knowledge and achievements, but there isn't an animal on the planet that would trade places with you and live in your cities away from the Earth, and with all your worries.

Living in the present makes a crucial difference that enables you to be fully alive. When your mind, body and soul are united you are in your full power, and very, very few of you manage this consistently at the moment. Allowing yourself to unify your energetic and physical bodies through meditation will vibrantly reconnect you to your higher self and will let you feel alive during the changes happening in the next few years. You are not used to how this will feel, to be so alive, and not all change is something to fear.

The direction you will find yourselves drawn to will be toward actively and purposely enjoying your lives. What happens on the way from your life today to that position will be the period of change. This will go from settled as you are now through many changes to a new settled life. The changes will have to occur to have a new life at the end of this period, and this is where the advice about letting go of everything with ease will help. It is easier to open your hand and let something drop than to have it prised out of your fingers. By the time your fingers have been prised apart everyone else will have walked ahead. And that's ok because this is not a race, and you are not all expected to finish together; but why would you want to be the last one to start enjoying your life?

What kind of changes might take place? First up is anything that is already crumbling as an institution or infrastructure, as the energy is already leaving and moving elsewhere. Is your life's work tied to such an institution? Maybe you can do the same job in a completely new way. Perhaps someone will ask you to join them in a new venture, and remembering about the energy and new growth of spring, you say yes. Your life changes because you are not dealing with the entropy of the old, but are in step with the planet and the prevailing energy. It's demoralising to use your own energy to prop up something that is half-way to falling down. Some institutions will be reformed and suddenly start working again as they were originally intended. You have a gap between what you expect from them and what you receive, and that also drains your own energy as you try to work out what's really going on. There are too many hidden aspects to large corporations and government departments, and too many lies; you waste your lives trying to guess how they are going to treat you. Maybe you like living like this, but you haven't tried living in an honest world. Truth is an aspect of light.

Has this happened every 26,000 years? It has never happened as dramatically as it will now, and that is because you are part of the picture. Your inability to see the higher dimensions allowed an extra-thick layer of darkness to accumulate here, and that energy supported some quite evil institutions around the globe. These will struggle to keep their power, but in the end they will fall. When you add these power vacuums to the other changes you will be dealing with in many areas you will have a lot of restructuring to do. There is also big of cleanup of toxic damage that humanity has wreaked on the planet which will be an ongoing project. You would not have lived this way if you had not been blindfolded. You are a soul of light and chose to play a risky game, but you cannot look at yourself in isolation any longer (we hope,) and will see that you are a piece of a larger whole. What you have learned on Earth is of value to us all.

What did the rest of the universe, from one side to the other, learn from humanity playing blind man's bluff on Earth? We learned how strong love can be when challenged, really challenged, by hopelessness and despair. You joined together in the last century to put people into concentration camps and prison camps where love and sacrifice survived. We cried at your bravery, and if humans had not built camps how would you know the depth and strength of love? This does not need to be repeated to illustrate that lesson again. On the other hand you also showed how loveless you could be, and how easily you throw away another's life. This seems to take place everywhere on your planet. We cry at that, too. We mentioned that this one game showed how far apart light and dark could be pushed and still be one experience, one game. You allowed these lessons to be learned, and carried the pain as a soul, but it is not a permanent burden. However, we don't see any other soul group volunteering to live this way.

You can only see the light against the background of dark. It is meaningless to paint a picture with white paint on a white canvass. You plunged into this game and painted a huge mural that we've all been busy looking at, and learning from your experiences. Because this was such an extreme game you removed whole swathes of games with minor variations that could have been played as experiments and therefore shortened the lifespan of the entire universe.

Why 2012 Was Always Going to be Different

    This universe is a replica of many, many universes that are busy playing games with varied outcomes. The universes are like the leaves on a large tree, each leaf being one universe. Your closest universes are the ones on your branch of the tree, and you recently began a process of aligning with them. If you picked leaves and laid them flat on top of each other they would stack up and it's the same with your local universes. If you look closely at these universes you can see that some have finished completely and are waiting to be drawn back into the Creator. One is a swishing bag of light, where the flow is unimpeded and all looks like white sand. It has ended as a universe of light. Another is dark, stagnant and cold with no movement or flow at all; it has ended as a universe of darkness. It doesn't matter to God how they end as all will be drawn back in to Him. It matters to the living beings inside, and it matters to us. We are game players and we look for this universe to end in love, not misery. We will keep promoting this until the final end. The fact that you have cut short a pathway of games is promising to us, if this hardest of all games ends in light then we feel good about our chances of this universe ending in light. That's one big reason why this human adventure on Earth has attracted so much attention in the wider universe.

    The second big reason for all the attention is the sheer number of souls the Earth collected together to reunite and advance to the light. She has orchestrated a little bit of returning to God in her part of the universe. Coming together is sharing love, staying apart means you have a longer road to travel to find your way back. You are going to be one of the largest united groups of light in this universe when you are all ready to ascend. We're excited about it, you will hold the vibration of love in a very big way.

    There are risks involved in being alive, sometimes you might get hurt, or your heart may get broken. This universe exists for one reason, to experience life and expand the knowledge of what it is to be alive. This information returns to the Creator of the universe and He can better know Himself by having personally participated in so many games. He has divided Himself into smaller beings that can live all of these lives. These then divide into say, people, and they have the tiniest lives of all and forget there is anything more to life than what is in front of them. All these countless experiences are part of one Creator.

On the day the combined souls of this planet ascend together you will have become one, and you will have rolled up a large number of souls into one gigantic soul of light. That is literally the beginning of the end for this universe. As a soul of light you will have tipped the balance towards light and unity inside this paper bag for the first time. You will be a large ball of brightness and light, but most importantly you will be combining your experiences and sharing them. This is what the very end of the universe will be, combining experiences and sharing them in full with the Creator. You will hold a larger piece of God as you are made of more soul groups, and feel more a part of the Creator while continuing to participate in the universe. This is why ascension involves being "God realised"— you realise you are a part of God. It will have been a long road for you, learning it the hard way by playing blind man's bluff.

All this information may be difficult to take in on one reading, so our suggestion is that you lie down in bed and make this statement before falling asleep: "I would like to understand more about what I just read." We will show and teach you overnight and during the day it will slowly bubble up from your subconscious. Over time its a good way to understand new thoughts and ideas. If you know and understand something on the inside, when you hear an idea or information that is wrong you will feel the falseness of it, and you will have a measure of truth in your subconscious for comparison. When you are lied to you will hear it clunk against the truth inside, and when you hear truth it will ring clearly. Truth is light and light does not clash with light. Then it is up to you to act on your gut feelings. Many people are afraid to act, not realising that others agree and will join with them. This is one of the main roles for light workers.

The most effective method of manipulating people is by lying to them. The general populace will answer opinion polls on the least trustworthy professions, and expect salespersons to lie in order to sell their products. This is humanity's way of tolerating lies, and you try to take evasive action as individuals. You may walk away from purchasing a second hand car if you think all is not right. Those of you who are ready to say "Stop lying to me" would do better to join together and aim it at the controlling liars that form all of your worldwide governments. We would like to be able to name a truthful government, but we can't. They work 85% in the shadows in the western democracies, and that percentage of hidden acts increases in the more totalitarian regimes. They are so few, and they control whole populations with outright lies coupled with hidden agendas.

This is terribly serious to us, that some of you try to live in the light, yet are living in societies that are made with building blocks of darkness. You have institutionalised darkness and everything you do is pushing against walls made of lies. The rules of law and the values of your societies are skewed to benefit only some and are extremely unfair to the majority. To step outside of society will make you an outcast, so you live and work in these unfavourable conditions. This is far from living where everyone has an even chance at life, at having enough food or shelter. These lies have made bars to a cage, and you are living inside trying to live lives of light. You who are beings of light have consented to live under their rules, and seldom protest. You can create change with only your voice and your written word, the groundswell of opinion will grow behind those few who have come here to take direct action. For every person who actively protests in person, there is needed a consensus of opinion behind them that supports their action. Change comes when the balance tips.

Perhaps by now you may be thinking that this will be solved by the new energy of 2012, but it is more correct that this good energy will help you to help yourselves. We encourage you to be brave and after you think the thoughts "they're lying about this" you need to say it out loud. Others will have come to the same conclusion. You are complicit[1] with liars if you permit them to lie to you without challenging them. All governments are there to serve their people and organise roads, schools and hospitals, etc. for common use; serving the people - not serving themselves or their friends. We can see the problem, but we cannot act on it, only encourage you to act on your own behalves. When you speak up it comes out from the shadows into the light where everyone can see the lie and deal with it. The best way to remove the secret dealings that take place in the shadows is to shine the light of truth and remove the shadows themselves.

What if you don't? There is energy in standing up for the light, and entropy in lying down and going with the status quo. We mentioned the flow present in the universes of light and the cold stagnation of the universes of darkness. Ending in cold and stagnation isn't something that happens to you from the outside, this is brought about by your own inactivity. You are here to be active, not observe, and if you consider yourself

---

[1] Complicit is used here in the legal sense; when a person stands by and watches a crime committed, even if he is not actively involved, he is complicit and jointly responsible as he made no effort to prevent the wrongdoing.

a light worker of any variety then pointing out lies may get the ball rolling. There was a reason why churches called Satan "the father of lies" as their symbol of evil. From the lies comes the structure of your modern societies, where life is lived according to false premises. You are all accustomed to living where you either know something is true, something is false but declared to be true to see if you will believe it, or cases where you just don't know. You will enjoy life more when you can trust people to tell you the truth. Work towards truth steadily through the year and the energy of 2012 will support you. You made a good start towards this in the previous year with the Occupy movements worldwide, and the Arab Spring. There will be more to do in the coming years.

7

The New Earth

AS LONG AGO as the writing of the New Testament of the Bible, the Apostle John had a vision of a New Earth coming to replace the old one. In John's time there was less pollution or global warming, but they knew the land had once been better, fresher somehow. This was important, and even then they knew the Earth's light had dimmed. They could feel it through the soles of their feet, and they could sense it in the way people behaved to each other. The old paradise of the Garden of Eden had been lost in the distant past and there was no going back into it to live again. The Revelations of John promised many wars and plagues, along with the reward of a new heaven and a new Earth. People have been waiting ever since for Armageddon and the end, without understanding what having a new Earth means for them. The mystic Apostle John saw much in his visions and knew they were true, as one does when one is given a vision. He reported what he saw but he did not interpret them fully, as they would only become clear when the events came to pass. He was a talented seer and we stand by what he witnessed, for to us it was just yesterday that we gave him the visions.

Where did the idea of a new Earth come from? You have had a long relationship with her and you compare the world you live in today to the original paradise world. And in your hearts you know this planet could be a paradise world again, but it has been lost over the ages under a scum of ugliness. It's not worn out; it's just struggling like someone gasping for

Why 2012 Was Always Going to be Different

breath, and it doesn't have enough vitality to remake itself as it once did. The planet is an energy being of light; it can take any physical shape it chooses and alter itself. It has had a variety of surfaces in the past, depending on what was required by those living on top, or under the ground. Your particular version of Earth is the most extreme in many ways as you have so many alive here at one time in a very complicated set of games. In the past she has been gaseous, she has lived with other planets in other solar systems, she's had time off as a soul of light with no physical presence, and she has been a water and ice world. Water and ice worlds host quite extreme games of flow and frozen stagnation.

Today there are still a few lovely unspoilt wilderness areas that are reminders of what the whole planet was once like. How are we going to get to live on a new Earth, all fresh and clean?

The other beings on Earth, in particular the elementals, love this planet with all their hearts. They laid their plans for a new Earth and as soon as she woke up in 2011 they requested elemental reinforcements. These were released from other less urgent jobs and came up to the surface to help. The Earth has been an extremely busy construction site starting with scaffolding around the trees. A year later there is an incredible structure being built far into the sky with beams going straight through houses. Elementals have been working steadily day and night at the task. The new Earth is already being built around us but we are underneath the structure with no view of the surface, and we are not used to looking into the higher dimensions. She will be ready by December 2012, and then this Earth will rise up to fit into the new surface, and the energetic new Earth will become the surface of the old. She'll be ready for another 26,000 years, if need be.

This new surface consists of energy for a beautiful planet,

but it does not have energy to support human cities. Humans create and support cities with their own energy, and they are not a requirement of human life. In cities you are collected into a large group of people creating fog and a lack of clarity, and of diversion from all the things that are really important to your lives. Cities don't make you happy; it's your relationships with other people that make you happy. There are only so many friends you have in cities regardless of how many live there.

Why ask the elementals to build a new planet if she is a powerful being and could easily remake herself in any form? If she really dissolved and remade herself many species would die, as your homes vanished. She has used different methods over past cycles to remake herself for new species, and in this cycle she is using the help of elementals. She is managing it all without the help of any of her guests. Elementals have a different relationship to Earth than we do; they are much more in tune with her. It is an honour for them to do this for her and they have been working with others in this great task in an unobtrusive way. There have been crystals helping in the preparations also, another life form close to her heart. Crystals have been the ones to design and adjust the blueprint the elementals have been working from and these two groups have been doing the actual work of remaking the Earth.

In December 2008 there was brought into existence on this planet the crystal with the human heart. He (most definitely a male) was foretold a long time ago as one whose birth would set the final countdown running. He learned about his crystal nature in the Cave of Crystal Elders, and he learned about humanity by being taken everywhere with his human mother. One day when he was big enough, and he grew fast, he retired under Silbury Hill in Wiltshire, England and fully connected into the Earth in her engine room. There he began designing the blueprint for the new Earth as a 3-D hologram. This would be laid over the Earth and she would rise up to fit it one day. The energy provided by the Central Sun will be enough to lift this planet into her new skin in an instant. December 12th or thereabouts would be a very good time to be out in the countryside with your feet on the ground and taking in the energy. It is almost complete at writing, and will be ready for December 2012. This is an event that will take place in an instant, but it's hard for us to tell you the exact moment. It depends on your time zone, and also we like to keep some information to ourselves. Focus on the entire year, not one day.

There has been a lot written about what people think will happen at the end of the Mayan calendar, but the Earth will go on and life will go on. Putting a new energy skin on a body changes the body, but energy comes first and physical change comes afterwards. December 2012 is when the old Earth will rise into the skin of the new Earth with a gentle sigh. This is an instant of renewal and only some of you will consciously feel the changes. We are talking about energy changes, not physical ones. This is the point when you have a lovely, new clean planet and you won't want someone's dirty oil spilled on the beaches, or pollution fouling the air. This will be unbearable and intolerable, and you will either have to live with this feeling or do something about it. The change in energy will support a change in direction in the way people live their lives, for those who desire to change.

The speed of change relies on how fast change can be absorbed. This is why we urge you to let go of everything you can and release yourselves to move forward with ease. We are not talking only about physical property and possessions, but also about old thought patterns. The old thought patterns are what drove you to burden yourselves with so many possessions in the first place. Old, stale thought patterns that you inherited from your parents are the last things you need in a new world, and jettisoning these will let you arrive faster into the new Earth that is waiting for you. There is no advantage to dragging your heels in discarding what is useless, and your loved ones will get there faster if the overall momentum is greater. Don't stop and look back and wait for others, let them catch up to you. They'll get there in the end, sucked along in the flow that is created. Letting go is a personal act, you can encourage others to let go of their stuff but you can't do it for them. Even if they never let go of anything, they will arrive many years later, and will have missed being there at the beginning.

What kind of world is on this blueprint? For one thing, there are no buildings in the middle of Avebury Stone Circle. That's no place for a village to be, it would be like living inside a large cathedral. The town will fall into disuse and eventually be gone, as people take on caring for the circle again. Cities will be smaller and more pleasant, pleasing to the eye and feel more balanced and healthy. Many people will continue to live in them, but more middle-sized towns will better serve their local communities People will make an effort to build lives that keep them out of the cities, and see others for social reasons rather than working alongside them in office buildings. Being shut up in a building and working for most of your lives at a desk will no longer happen. Many of you have spent your lives working like this already, and you have helped build a future where no one will have to do that anymore. It's not healthy to be apart from the Earth. Work will be more varied and of smaller scale, with an end to commuting. If an activity doesn't promote human or planetary health, you won't waste your time doing it.

These are quite large changes, and will involve a number of transitional years. The rate of change will be variable and not always predictable. The main thing to remember is that when the energy has changed, the physical will follow, and that will be a result of humanity's own changes. Your cities won't be abandoned overnight, but you may leave them to live in closer contact with the Earth. You will want to live closer to her, and will feel too cut off if you can't feel her under your feet. You are connected to her, and will wish to benefit from feeling that connection of love. This is part of what has been missing all these years for humanity, and you are ready to resume your roles in a learning partnership with her. The rate of change will be fastest where people let go of their burdens and move forward, resulting in areas changing at different rates. Be happy and welcome the new world into your life.

8

Moving From the Old Earth to the New

LET'S RETURN to what you already know about the new Earth. This has been written about many times by your fictional writers, because to present the information in fiction is a way to get people to accept new possibilities. The authors create a different world, and by writing about it open up the possibility of its existence. If the readers can imagine it, then they can help create it by admitting it is possible; once the idea is public it is "out there" and has a place in the human consciousness. This is why you need to be careful what you give your energy to, because you would not want to meet all your creations in the real world. A good example of creating something new is Roger Bannister, the first man to be timed running a mile in under four minutes. Men tried for years to run a mile faster than four minutes, but no one succeeded. In 1954 Bannister ran a four minute mile and suddenly lots of men started running that fast. This was to do with creating the possibility in everyone's minds of running the mile that quickly. There are many more examples of mass learning among humans and animals.

The lands of Narnia by C S Lewis and Middle Earth by J R R Tolkien were the closest to the real Earth. What was so real about them? They captured the reality of many different species working together; and escaped from the fiction that man was the only one who mattered. Included were conscious trees and lands of great beauty. They showed man joining in common cause against the enemies of the land and fighting for it. It was fiction, and also a model for working happily together. Narnia in particular showed a world where all animals communicated and worked together in their lives, they joined with fawns, centaurs and others we no longer see to make a many-layered weaving of lives. When left alone without an evil overlord they were simply happy. They didn't achieve much beyond joyful living and food and shelter, and they wanted nothing else. Interestingly there are books in the series with non-Narnian animals, and they live as if part of them were missing. I think we're shown those other animals as a contrast between being conscious of life as we go through the world, and walking with our eyes shut.

The Last Battle by C S Lewis, the final Narnia book, describes the creation of the new Earth. It is a world like the old Narnia with the same landmarks but everything is more alive and fresh, and newly made. The old Narnia falls away until only the new remains. Not everyone chooses to come to the new world to live, some refuse to believe it's happening before their eyes and they stay behind. One doorway leads to the new Narnia, or it leads to wherever one believes it will lead, and some see a horrible future there. Only those that go willingly through the doorway move forward into the new world. Others sit in the dirt while the opportunity passes by. Some people will not be ready to come with you when you pass through the doorway to the new Earth, and where does that leave you? You will have the choice of bravely moving into the unknown and engaging with the real world, or you can sit tight and wonder why everything is starting to change around you. It's about how easy can you make this change for yourself. If you want someone to come through the doorway with you the best way is to go forward yourself and lead the way. In The Last Battle the dark days are over and the days of joy begin, and they are never going to be separated from the love of Aslan again. Their world is full of new life, and no one wishes for the return of the old world.

Do you know why we wanted C S Lewis to write this book? Because sometimes fear can hold you back from walking forward to something that you will love and enjoy. We don't want anyone to be afraid to accept good things for themselves, or fear to actively seek out the best. We wanted you to know that you have a continuing life on this planet in partnership with her, and that this life will be familiar and acceptable. She isn't going to remake herself into something completely different like a water world. She will still be the Earth, and you will still be you. You will feel different and begin to alter your lives and have a period of adjustment, but the Earth is not looking for the destruction of the human race, that would be pointless now. She is so close to achieving what she set out to do that she wishes to take everyone with her if she can. We want you to look forward to the love and light and warmth of what's coming for you all.

At the moment the old and new Earths are present, flickering in and out of view. One Earth is here in the lower three dimensions, and one is being readied in the higher dimensions. The one in the lower dimensions will not go forward into the future, it's time will be over and the two Earths will combine using the energy of the alignment with the Central Sun. The attributes of the old Earth will drop away once the switchover into the new Earth has happened. It's happened many times in the past, here and elsewhere. This is the method that causes the least disruption to the resident life forms, and is the gentlest to live through. This is not an occasion when many people will die due to the changeover, but it will affect everyone alive. When the Earth shifts into its new body everything under your feet will have changed. This will affect all of life; the animals and plants know it is coming and are already rejoicing, and many people will be bewildered. Their collective subconscious will know they are walking on a different planet than before and it is not the same. And that's when the fun begins, that's when the light lays bare what others desire to remain hidden. You will see man-made institutions crumble and release their power over you. It will be a relief to so many of you and others will feel dismay when they recognise their role in what has been happening. Everyone can choose to change.

There will be some immediate changes that you may become aware of, and there are some animals that will move on as a species. These will be fewer than in the past, as most are staying on for the unified ascension, nevertheless some will be leaving and you will miss their presence. These animals are not the ones that humans are driving into extinction. There will be some people that just don't fit in energetically, and they will unexpectedly pass away. Here and there you will find people dying who are not reaching old age but are choosing to leave. They will have finished their lives for the time being, and they will incarnate again one day when the time is right. We know that dying is a most natural event, there is nothing to worry about when someone dies, and everyone will die one day. Death is not something to be afraid of. However there are those who will not cope with the new Earth, it will be too confusing or they won't want to face the way they acted in the past. The rest of you will have time to change, and choose a new direction because you no longer wish to live in the old way. What we want to get across is that change is coming, but not coming overnight. First the planet changes, then you begin to adjust to the new Earth, the new energy and the new experience of having the universe become familiar to you once more.

Downloading information is a familiar term from the internet, and now you will be downloading information from the organic universe into the matrix of a living planet. There is a lot of information out there that is shared, and you are now going to get glimpses of it in your dreams to start with, and inspirational ideas in waking hours. You have been ignorant of technology in common use on other worlds, and more importantly there is a shared knowledge of what it means to be alive that you haven't had the opportunity to grasp. Because the reason you are alive is a mystery you have formed theories and hypotheses in explanation, and then you have turned these into a number of religions. These religions have then striven to dominate as many people as possible and become powerful, and fight other religions. It's a long and sad history going back millennia, and we don't know if it is an excuse to fight others, or a way of differentiating others from yourselves. We don't see killing people as an act of love or light and therefore see the institutions of religion as very dark and full of fear. They serve many purposes, but they do not bring you closer to God because they separate you from one another.

Does this mean that the new Earth will not have religions? They will be less compatible with the new energy, some will fade and others will reach a peak of strength before falling away. This will be part of the readjustment period. Those people who have something to gain through religion will be reluctant to let go of it, and will exploit it during the transition period. Remember you will be going through a period of change and religions are fundamentally conservative, and that means to preserve the status quo. Fear of change will send some running to their religions, but when change is subsiding and a new stability is in place they will lose their grip on their members. You will be left with the leaders who are there for power or money. But where these many religions have existed for centuries, they will decrease quite quickly once the Earth is renewed and a great divider of the populace will have disappeared. The final goal for humanity is to remember that they are one soul, and it is a part of God. Religions often require you to have an intermediary to speak to God, and insist you are not God yourself. Therefore we see religion as a blight upon your path to reuniting as one soul. Fortunately the energy of the New Earth will support you in remembering that you are all one.

9

A Society Created by Lies

ONE OF THE coming changes is that societies will have fewer boundaries and divisions. You have fences around fields, and borders around countries with passports and immigration controls. It will seem so unnecessary to have these artificial boundaries. You will be more aware of the limitations this brings and how it stops the flow of animals and people, and it will feel out of step with everything else. You may fear that everyone will come to your country and not leave, but that is not flow. We're talking about circulation and movement, and a planet that is able to bring rain to places where it is needed. The numbers of people who are refugees because of famine will be eased, and you will be left with those who run from vicious governments. The energy to support those regimes will have changed and they will not last many more years. Life will settle down again and the flow of people will be intuitive, where people migrate to the areas they need to be for a time, and then migrate again. You were once this kind of society, relaxed and not stuck for a lifetime in the wrong place. People know now when they are ready to move on, but they are not always allowed entry to new countries. It's about matching the person and the land under their feet.

It will begin to dawn on people that owning land and fencing out other species and people contains a fundamental flaw: you cannot claim to own a little portion of a living being. It's because you have no conscious relationship with the Earth that you say you own land. Native American Indians migrated from season to season living in touch with her; they lived on her bounty, but did not own her. This is the difference between truth and what is supported by your current societies and laws. You join together and support land ownership by living in accordance with the laws of property. Your laws are artificial and self-imposed, and there is no reason they cannot be rewritten. Remember this, because as the energy changes you will feel that you are in conflict with the way things should be and want to align your lives with what is true, not false. There will be a busy period where laws are changed and people will reject owning so much property, more than they can ever use in a lifetime. When they shed their ownership of the land you will begin to unite as one species. You simply cannot become one ascended soul when some own too much and some too little. The ownership of land gets in your way as it is the cornerstone of your laws of property.

You will not need revolution and riots to change the laws, but they may happen. There were riots in 2011 and those boys and girls are here because they are also your best people and were given the chance to incarnate here on Earth. Other souls stood aside and let them come through. If their riots begin a process of change they will have done something useful for everyone and played a role in the human story. They responded to the energy change, they are beings of light and deserve your recognition as such. You do not have to riot yourselves, but you can speak up and challenge in your own way lies and the outcomes of lies. This is the point where we want to say that everyone needs to be held accountable for their actions and lack of action. Somehow the thread that links action and accountability has been snapped in your societies for some. For those of you who take responsibility for your actions, why do you put up with those who don't? They get away with it because you allow them to, and in those cases their actions are lies. When you refuse to be manipulated by lies that final stumbling block of change will dissolve.

We said earlier that change would come in response to the new energy, and that there would be different rates of change right across the board. Refusing to allow yourselves to be lied to is the single biggest change, lies are at the core of all other problems. They allow all the rest to happen, and the entire house of cards will fall when there is enough light to expose all the lies. You can only act for yourself, and by acting even a little bit you will bring about change, and you won't be alone in your actions. Specifically we are talking about those who are deceiving others and have something to hide, hidden agendas, hidden accounting practices, hidden power. We've said it before, but your politeness at not challenging these people lets them rob you of a happier society. They are parasites, and they exist because you let them.

Lies create an artificial world around you, a world that is set up to benefit those who are telling the lies, and to disadvantage those who live in the world they created. All of you now alive are living in societies manufactured years ago to benefit some men who are long dead, but the laws they put in place are still controlling you. The greatest abuse centres on the laws of personal property, and on a planet where there is enough for everyone, even seven billion of you, too many do not have enough to eat. Others have more money in bank accounts than they could spend in a series of lifetimes. This is known and permitted by all of you. You feel you cannot challenge those in power but some very advanced souls have already shown you how to make peaceful and lasting changes; i.e. Nelson Mandela and the Mahatma Ghandi from your last century.

10

The Golden Wave of Energy

A LONG TIME ago the Earth was strong, and she fought off beings of darkness that wanted to move in and set up home. She had to turn a blind eye to these entities as part of the contract she agreed with humanity to host your game, so you could be given your chance to find God starting from complete forgetfulness with whatever risks that entailed. If she had banned these beings you would not have had your full opportunity to learn. As a result she quickly became overrun and plundered. They built nests under her surface and bred, and she became coated with tar created by your fear and unhappiness as they created their type of world. They settled in and used you as a free source of energy for as long as they could get you to produce the food they liked: fear. The Earth became like a Wild West town with no sheriff, and was overrun by the bad guys. We, the angels of light, looked after her as best we could. Some good and strong peoples like the aboriginals of Australia and Native American Indians looked after her, tending her like gardeners. Those were the only type of societies that remembered to fulfil this part of your contract here.

What's happening at the moment is that these remaining "primitive" societies are declining and as their numbers reduce they are less effective at helping the Earth. To try to get help from the rest of the population we have found and fast-tracked those who can hear us, to start up practices that will heal the Earth. This is why Planet Earth Today spent time explaining how Reiki Earth healing circles functioned in Atlantis. Today there are a few healing circles that take place regularly of many different healing types. They could be Shamanic, Reiki, Sufi spinning, or even a Scottish ceilidh dance where the laughter of the dancers and the lively music clears the energy. The result is the same; the beings of darkness flee the light. The areas these occur in are cleaner, and the people living there have the chance to have clearer heads and keep more of their energy for their own use. The Earth underlying them has been cleared and she can use those places as breathing holes. People in healing circles like these have helped keep the Earth alive for thousands of years. (Planets, like anyone alive, can die.) Those who partake and give freely of their time to help the Earth are blessed by the light. Using your attunements into Reiki, training as a shaman or Sufism for the benefit of the planet is healing for yourself at the same time. Treating the Earth as a patient is a big job and it stretches your abilities as a healer. She knows her friends and remembers them.

Earth healers are needed in all corners of the globe. We wrote about the method of Reiki healing parties where the energy is held on the outside edge while people in the centre talk, dance, eat and sing. Silence is not a requirement of healing, but enjoyment is.

What is the future of these healing groups? If the Earth is awake won't she start protecting herself again? She is waiting for the end of your game before she takes over her own protection and needs as many healing groups as people are willing to set up. As people have less scum surrounding them they will find it easier to see the higher dimensions that are already visible to some. Everything interlocks and everything matters, and this will help put you in the right place for you to identify the truth and speak out about it. There will be less fear, and you will feel braver. Humans are wonderful, and we know that you are not the crushed soul that you sometimes appear to be. You are beings of light yourselves and have suffered along with the Earth by living under a layer of tar. Healing groups are a means of connection with the planet under your feet, a way of showing her that you care about her well-being. The need for these groups will continue as long as you are resident here on Earth. This is not the only planet that lives with souls who provide healing in return for all they receive. You have a role on this planet that places you into a web of relationships, and your relationship with the Earth is the single most important one you have as a species.

At this point you have the ability to see five dimensions, you are all capable of it and it is part of being human. The vast majority don't believe there are more than three dimensions and like to laugh at anyone who can see more than they do. Seeing energy is a quick way to know who is ill as their energy is weaker in some places, or which seeds don't have enough life force to grow into healthy plants. Not that many centuries ago people could see elementals and talk to them, especially the ones who lived around their homes. Dragons and unicorns were spotted in the past and are remembered in your fairy tales, and they live on in reality in the higher dimensions.

This isn't about striving to see fairies; this is about knowing you can see them and catching their movement out of the corner of your eye. You can learn to see them by practising the same way as any new skill. It's all in front of your eyes, and you are allowing denial and other people's scorn to blind you. Acceptance will open a world of movement in front of you that will begin to fill in with colour and detail over time. By next year this will be even easier as there will be less blanketing darkness to hinder you. In addition to elementals, you can now see the ebbs and flows of energy surrounding plants, animals, people and the Earth itself. There are people who can walk to a place on the Earth and say "Feel the energy here. It feels like a ley line." They are using their senses, and you have the same senses. We want you to work on this, feeling and then seeing the energy of all living things. There could be nothing more natural for you than seeing the world you live in, and all who live here.

The Earth has flows and channels of energy, some are positive and some are negative. The polarity of this is part of your game of positives and negatives. The opposite of this would be a completely blended planet with no channels of energy. You requested complete polarity when the parameters of the game were set up to make it even harder for yourselves. The archaeological work on ancient sites in the Shetland and Orkney Islands reveal that homes were only built on sites of positive energy, and any negative areas were saved for the streets that ran between. Building your homes on negative sites does not support your health. Very few of you have any idea if you live or sleep over a negative spot on the Earth. This type of energy, sometimes known as geopathic stress is actually visible if you look for it, and it's possible to feel it as you walk across one of these streams of energy. Humans live more healthily over the positive areas.

If you can learn to see the positive and negative energy under the surface of the Earth, it will help you to spot it in the other places it exists here. It exists in the hearts of many people and not all acts of violence are inspired by dark angels. Humanity is also made up of light and dark. You need to use all your available senses to avoid those who hurt others without a second thought. You run your own lives, and can choose how to live. We guide, we advise and you decide how you wish to act. We are not the ones incarnate here making choices, we are here to help when you allow us to. Reading people's hearts is a matter of energy; do they contain love and light, or are they dark and vicious? Most are swaying somewhere in between, with only superficial knowledge of the world around them. It's your choice how you act, and who you spend your time with. We see people become victims to someone who has no love or kindness in their heart. It is possible at your stage of development to see this energetically before getting too close when you meet the wrong person. Sometimes it's referred to as following your instinct.

How does a soul of light who is blind to energy ever hope to finish their game and prepare for ascension? This will be possible when some lead the way and others follow their example. How does a soul of light behave to set an example? They act with love and joy, and others notice and are drawn to them because the underlying energy has changed for the better. You have been given many good examples and wise teachers over the millennia and the meaning of their words were eventually distorted. Religions were formed around them that promoted separation, isolation and poverty of spirit. Hatred was condoned as being justified in some way by the founding teacher. We are one hundred percent against religions for the pernicious effects on humanity. By preaching hatred and their lack of love and joy they show exactly the energy they carry. Shun them in all their forms. No one needs religion to reach God, or to know their own heart, and in their hearts they will find that spark of God that is present in all things. They are a plague on humanity. You are individually better than they would have you believe, and have the knowledge within you to recognise God. You don't need them to help you, but they need you or they are nothing.

2012 is the year that the universe provides a wave of golden light that travels outward from the Creator of us all. It has been timed to reach Earth during this year because it will carry the energy of a fresh start to help you initiate new actions. It will pick you up and set you down further along your road, and how much further is entirely up to you. We are hoping that many of you were ready for it and took a large jump forward on its energy. 2012 was designed carefully to help you move and flow, and the first event was January 1st when the energy changed. The old female energy that formed 2011 faded out and the bouncy new male energy of 2012 arrived. We had a few weeks of planning how to best use this year and then Mars retrograde arrived for more than two months. This period allowed you a breather, a time to work steadily without struggling to stay upright in the stiff breezes of energy of the first few weeks. When Mars turned direct in April humanity had had their time to prepare for the next wave of energy and this was the one you were waiting for, the tsunami to carry you forward and place you in exactly the right place to walk into that new world. This was a gift to all here on Earth.

The wave of energy was devised by the Creator to comfort and help the Earth. It was set in motion using the portal of the Central Sun and originated like a breath from the mouth of God. A breath of love and cherishing, this is the nature of the wave of light in 2012. The wave will be available to all on the planet, and it will be interesting to see what changes take place. Animals and insects, trees and plants will have this reminder that God is love, and that love is as strong as hate. You have all lived here for so long under the blanket of fear that drives hatred, having it washed away from all surfaces will change the energy of the planet, of everything except the insides of your hearts. From this point on you will be playing a game that involves your own soul and humanity will be back in charge of their own progress.

The Creator is planning on intervening on the side of love, in a game that you arranged yourselves and left you vulnerable to all the dark forces of the universe? We wrote in *The Downfall of Atlantis* about this contract between you and the Earth, about the underlying twist in the conditions that you agreed to: conditions that put you at a complete disadvantage for the duration of your game. Your human soul is unique, you have your own characteristics that are not repeated in any other soul in this universe. Souls are not just large, vague balls of light or grey. They are made up of personality, and part of your personality is that you are highly creative. We can suggest something, and then you take the germ of the idea and create something amazing. You also like to be challenged, so you created this extremely difficult game here. Think about it, the Creator sets this universe in motion with different soul groups, and you are one of the strands of Him that is creative. You are lively and fun to watch, we never know what you are going to do next. This time you were tricked, and your good faith and pride may have played a part there. However you could not go any further in this game with the original conditions in place. This planet and everything on it is going to be scrubbed clean, and if you fail to ascend to light it will be down to you. You will not be at the mercy of invisible foes. This is why we want you to look at the hearts of others before trusting them, as humanity will be running their own lives here after this.

When you began your game on Earth it was not to see how much money you could pile up, it was to see if you could find God from a position of complete unknowing. If this is achieved by your actions in life, and by the way you actively seek to live in love then you were on the right track. By understanding your relationship with God you can turn around to show others and they will begin to join you. The place they will be joining you is the New Earth. This is how change happens, person by person when the circumstances are right. There are many who are alive now who will never understand in this lifetime their relationship to God, but they won't live forever. These will eventually die out to be replaced by gentler and wiser souls of greater understanding. Different generations have been brought up in different circumstances and they embody the energy and influences of their times. The generation that was born after 1986 is carrying the future of the planet in their bodies, and they can be trusted to walk directly towards it and understand it. It will be safe with them.

The year 2012 was never going to be the same because of this universal wave of energy in April, and it's after effects. If you held the planet still and washed it with a fire hose it would be clean, and the spring tidal wave of energy will be doing exactly that. Suddenly, for some of you, the sludge you have been wading through will have vanished and you can walk freely. You may have felt that everything you tried to get organised and set in motion hit so many obstacles in the past, and never realised that you were exhausted by trying to push boulders up a hill. This cleaning is to benefit you, a valued soul of light, and is helped by the greater universal beings of light. There will be clean ground on which to go forward, and we feel that many of you will take off at a run.

Part Three
Life after 2012

11

Get Ready to be Happy

WHAT CAN WE expect to happen in the years following the rebirth of the Earth? Is there a golden age guaranteed, or will it be chaos? Some of what will happen in the future depends on the present, and preparation in 2012 will advance you to a better starting place for the future, just as preparations always do!

First we want to point out that the course of your human life is not set in concrete. Part of the reason for showing you the rewards of moving forward and the pitfalls of inertia is to encourage you to keep progressing. Spring, summer and autumn of 2012 are the periods of greatest flux on this planet that you will have ever experienced. From our perspective it looks like a battle zone with certain people striving to move forward and let go of everything that no longer serves them, while others throw obstacles in their way. It takes single-mindedness to walk past obstacles and not let them slow you down; you can however actually walk through them as if they don't exist.

Who would want to prevent you reaching your full potential as a soul group? The dark angels of the universe have long been resident on Earth and wanted your planet as a prize. They gain nothing by Earth's and humanity's ascension to light, as it will be the end of their hold on her. If she is able to break free of them in spite of everything she has been through, she will lead other planets by example. You don't know how important this one act of ascension will be, how it will echo across the universe as a model for all other planets. Those planets who have played it safe under their planetary shields may reconsider hosting new games. Her multi-soul ascension will be taking an active step for the light.

We speak about love as if you all know what we mean, but some humans have not yet experienced love in their lives. Think about whom those people might be, and we're not simply talking here about war orphans. It can be ice-cold adults who have married and raised families. Love is warmth and happiness, true joy in your surroundings, your friends and families, your work, and your daily lives. It can radiate outwards from those who hold it in their hearts, just as the ice-cold parent emits no warmth. The lack of love is outwardly expressed in posture, face and voice and others know when it is absent, while some show their hearts easily. The source of all love is God, in quantities far greater than you can experience as a tiny physical person, but still you hold as much as you are able. Jesus is a good example of one who held in his person so much love that others crowded around him to experience just what love could feel like. God is also the source of all hate, as everything is part of God and it existed as part of Him before He made this universe.

Love makes you happy and joyful, and joy is another form of light. We really don't see much love and joy when we look around. The most joyful humans are the youngest, the babies who are adored by their families; and rightly so. As they giggle and laugh their families laugh and smile and joy is shared by all of them for a time. Precious babies and their gift of unadulterated joy; it soon begins to dim. Toddlers laugh, young children laugh but there comes a point when children start to look worried if they laugh in public. They've been told to keep quiet and feel uncomfortable just laughing out in a relaxed manner. Someone will make them feel bad about being happy if they do, perhaps the 'cool' set in school. You know all this; you've lived through it. You are missing light in the form of joy from your lives. Would you like it to return? We hate to say practice being loving and joyful, but practice laughing out loud without restraint and observe what happens around you. Others will join in, not everyone, but joyful laughter is infectious and it spreads. You won't be laughing alone, and why should you care if you are the only person who is happy? If misery loves company let them find someone else.

We see so little genuine happiness that we're not sure you will necessarily know what it feels like to be happy. You may feel that you are experiencing something unfamiliar and maybe not exactly right if you let yourself be happy. For many reasons there are a lot of unhappy and depressed people around and they get accustomed to feeling that way. We actually feel that happiness is going to be harder to adjust to than love because while people will say "I'm single now and looking for a partner and love." Very few say "I'm habitually sad and never feel happy about anything." But we can see this blanket of sadness and worry when we look at you. Cities are the unhappiest, where those who live there are isolated from the touch of the Earth.

Why do you all worry about the future? What do you personally gain by worrying? Does it fix the future for you, or spoil your enjoyment of the present? The worry is based on a lack of trust that the universe will look after you. Going back to one of the other universes that exists as a swishing bag of light, what would happen there if the particles in free-flow started to worry about where they swished to next? They would begin to slow down and darken. Worry slows down your ability to flow with ease from one situation to the next. You begin to attach emotional baggage to whatever you are worrying about, and you may start to be afraid that it will turn out badly. These are not feelings that make you feel light and happy. We can see you accumulate darkness around you as you worry until you can't see your way through it to find a way out. Trusting the universe actually means that any result is good because you flowed into it on a cushion of light, rather than stumble into it in an impenetrable fog. That way you bring the old problems with you into the new situation.

If you flow into a new situation, what do you bring to it? You bring love and joy, and these are the energies you will be working with. This is where humanity has hit its absolute peaks in the past; those people who enter into the most dreadful situations with their light shining until the end. If some of you could maintain that level of light through death camps as beacons of brightness (and you did) then anything is possible. Love and joy are stronger than the deepest despair. Some of those people refused to despair, and as you live your life, do you want to live it happy or miserable? In your own lives, if you lose your job you can live the rest of your lives happy or sad, and if worry makes you sad and fearful you need to find a way to break the pattern. We like "Just for today, do not worry" one of the five Reiki principles taught by Mikao Usui. It begins to break the habit of worry and darkness. The more darkness you generate the harder it is for the light to reach you, and for the light to shine across this planet. There are plenty of dark angels who will use the fear and despair as food.

We talked about love and joy equalling light, but we want to make sure you know that darkness is made up out of fear, worry, misery and despair. Sometimes you can walk into an area that seems to hold these energies, and know you didn't feel that way before you arrived there. Funeral homes and crematoriums, hospitals and some churches may contain these energies. Or you can work up a good batch all by yourself. If that is the case you need to be aware of what you are doing and train yourself into new habits. Gratitude is a good antidote to many of these vibrations. Gratitude is something you do for yourself, others never get as much out of it as you do by expressing gratitude. It's a transitional vibration leading you away from darkness towards light.

You are going to have a new planet that will start off clean energetically. We wrote about the energies of light and dark so that you would understand your role in creating these vibrations. You choose what you create, and your choices have been formed by the societies and families you were brought up in. We are saying you are free to choose something else; and where there was a habit of worry you can have a new habit of joy. Joy is not going to be given to you because your planet is renewed, but it should be easier when there is joy and light coming through your feet to start to share that feeling. You will have some help from the planet being joyful but we are asking you to be aware of your own contribution to the energy balance of the planet. We want you to live lives of love and joy and blissful happiness because everyone deserves that!

12

Changing Your Lives

Following December 2012 you may feel an incredible relief that after a lifetime of living on a planet where you may have felt the odd one out, everything now feels very right. It depends on you and how you are able to feel the energy surrounding you, but there will be those who are comfortable for the very first time. It's stressful to be a person of light surrounded by darkness because one has to be strong to live with those who are holding the opposite energy. Some of these others will wake up and look around after sleepwalking through their lives. This will be in response to the Earth and the fog of darkness being burned away. They'll be pleased at the new Earth but they won't understand what has happened. Others will not understand, and still others will resist and start spreading lies and fear again. You have a very good window of opportunity here for anchoring light, for refuting lies, for spreading joy. This window will vary according to how much light is generated by humans, but you should have a few years at least.

What steps do you need to take in your own lives to be happy? This is one of your top priorities. It is also a wonderful time to plan some events just for the fun of it. Parties, picnics, music, theatre, hikes, retreats, whatever you feel will put a smile on your face. The energy of the New Earth will support events with these vibrations, and you will be supporting her. We are hoping many of you will let go of drudgery and chores and find simpler and easier ways to live. Light workers, this is when you can plan healing activities and find they are in demand. Put dates on a calendar and plans into motion, and trust that all will be well. You will be carrying the vibration of the New Earth, and others will want to know what it is all about. Even those who did not immediately notice will start to realise there is something very big happening and they will want to learn about it. As always, some will not notice in this lifetime.

You personally will be able to contribute to this universe ending in light and joy by your own actions over the next decade. This is why you incarnated here and now because all souls and beings in the universe want to matter, and make a difference. You will be able to spread the vibration of happiness and change other people's points of view. Some of you may feel like that is calling for more bravery than you currently possess, to be a shining example to others. You didn't incarnate to step back and give up, but to take a deep breath and say things like "I really notice the difference since 2012, I feel like the Earth has changed. It feels wonderful." "I'm enjoying my life." The other hard step could be that you are currently not enjoying your lives and you need to change something fundamental in the way you are living. You may have the wrong job, the wrong partner, and the wrong location to be really happy. It requires an absence of fear to make the big changes in your lives, and trust that you will be looked after once you make those changes.

When you have made the changes you need to be happy, then you are anchoring light in the form of joy. Others will notice that you seem to be so happy all the time now, and want to copy you. This is exactly what we mean when we write about one person taking a step at a time, and others following. After all the lives and learning experiences you've had, and having reached the point where you find yourself reading this book, are you content to say "I'll stick with unhappiness, thank you. It seems easier." There is a difference between drifting through life vacantly, and flowing towards the light. Flowing and experiencing everything that is put in your path is living, and drifting is not acknowledging the lessons that every person and event brings to you. That's like sleepwalking. There may be a planet full of those who exist by taking the easiest option in every case, but they aren't reading this book. You are, and we are here to support you in your life, and we want everyone to be brave and enjoy their life.

We keep mentioning that some will not notice, and it's not a device to prove we're never wrong. Until their dying day there will be some who think everyone else is crazy and nothing ever changed. Don't worry about these people, and do not waste your time trying to convince them. You'll soon know if you're talking to one of them. They could drain you of energy in the effort you make to convince them that the world has changed, and that would be your mistake in allowing them. In the New Earth you will not have to waste your time doing this any longer as there will be many who are on your same wavelength. These less sensitive people are part of the great human soul and you must never look down on them, as they are a part of yourself.

There will be great adjustments and changes taking place during the first transitional years after 2012. There will be upheavals similar to the Arab Spring where the people found they were strong enough to throw off oppressive regimes, or correcting the world financial systems. There is a place for banks, and for loans to help people start or expand their businesses or buy houses. You do not need to jettison all bankers or money lending, but something else has attached itself energetically to these structures, and that is greed and lust for money. The banks themselves could only see columns of figures and profits for themselves, and money for their personal salaries taken from their clients. As we said before, more money than any of them could spend in a lifetime. Only when it emerged from their hidden accounting practices into the light did it become clear how much money was being siphoned from the many to the few, and changes were made. As the public protested, a few backed down. What enabled protest was seeing the truth about extent of the greed, and how far sunk into avarice the economic ruling elite had become. Centuries earlier the wealthy were the world's philanthropists, they could see no need to keep all the money for themselves. That has changed.

The banking scandals emerged as the banks toppled the world economies through greed, and they were the only ones who emerged unscathed and with their lives intact. In the UK the banks passed into public ownership and salaries were paid by taxpayers. This allowed the bonuses to finally be queried in a way the shareholders had failed to do in the past. It became a matter of public outrage, and hidden no longer, how many millions per year these thousands of bankers and other businessmen were paid. It also became clear that they did not have a link between their actions and rewards, receiving bonuses for destroying business because the bonus was in their contract. There is a universal truth that you have to take responsibility for your actions.

When you are responsible for an action, you are responsible for its consequences. This is how this universe was set up so that beings could learn and progress. What do you learn if you are not responsible for your actions? You learn you can get away with anything. When we mentioned that the human game had gone as far as it could in one direction it is exactly this point we had in mind. You can go no further towards learning unless you learn from your mistakes. Once the consequences are removed from your actions and you stop learning you have reached a dead end. As humanity turns back now from this position there is the potential for extreme disruption. The way to lessen this is to hold yourself to account for your own actions, and speak up and challenge politicians and bankers who are acting without responsibility. The entire society could turn around at once or have a more gradual swing back into the other direction as some of you lead the way. Then the shock will be less.

Picture the Earth going through her rebirth, but alert now on the inside and trying to provide you with a steady surface. Now picture humanity on top in convulsions and turmoil because they are having secrets exposed by the light. This is like lancing an infected boil; it could hurt. You are going to live through this period of change and it looks painful, but it may be more painful for some than for others. We can clearly see that there will be many who have no pain and no hard times, but immediately create lives of great joy. Free from lies and living without a coating of darkness they are moving forward swiftly and setting up new business ventures, new homes and new families. They were just waiting for the obstacles to disappear. Who are these people?

Those who will slide painlessly into happy circumstances are those who have let go of their attachments to everything in their lives. They have homes and jobs now, but they will have homes and work to do in the future. They don't hang on one day longer than is necessary, but allow their grasp to loosen and things to fall away easily as they arrive at the next event or item. Remember flow is light, and stagnation is darkness. If you were floating on a beautiful swift stream, and were trying to stop all movement by hanging onto the grass at the edges as hard as you could, (that is how we see many of your lives right now); as if you will never be able to float any where else as nice. If you released your strenuous efforts to hold on and flowed with this stream of energy it would bring you to new places with new gifts for you. You wouldn't even have to swim there or look for them. This is flow, and this is the flow of universal energy that is always present. We are very sad sometimes when we see the fear that makes you hold on so tight.

Those of you who let go and trust that universal energy will take you where you need to go will change your own personal energy, and that will change the energy of those around you. Those hanging on will maintain their rigid fear and that will affect others around them. It's something you always carry; your own personal energy field. As more of you trust and are able to let go, there will be a knock-on affect on whole areas, invisibly influencing others just by moving among them. Let go of everything and don't bother to look back, you will only see a pile of belongings you have outgrown and discarded.

Back to the Earth with the human race in turmoil, now can you see that if you let light into your personal lives you will spread calmness and truth around you, and show others that there are alternatives to worry and pain. You will diminish the turmoil in your personal lives and surroundings. This is how many of you will skate over the surface of scandals and restructuring. Others will learn in the end that hanging onto the grass at the edge of a swift stream can't be kept up forever. As you let go and flow to the good things the universe is waiting to give you there will be no regrets, and those things you discarded are never thought about again. You never needed to burden yourself with so many beliefs and possessions in the first place, so many of the beliefs were untrue. As the turmoil among humanity lessens it gets easier for all the other species and the planet to maintain their equilibrium.

We are looking at five or ten years of humanity sorting itself out and getting rid of the puss that poured from that boil. Although we mentioned politicians and bankers, there are so many more hidden areas of activity that will be laid bare. At the end of these years you will have reinstated fair government and business practices, and religion will have slunk away in disgrace. There are truths about religion that will also be exposed in those years. Do not turn away from the findings that will emerge, but support justice being done. In the Roman Catholic Church they turned a blind eye to the pain of all the children they abused. This is well-known but there has been only a little action to challenge them to date. This is the kind of hidden activity that will be exposed for all eyes to see. It will be cathartic for all those abused children, now adults, to have their experiences acknowledged. The Catholic Church is used here as an example, but there are abuses of trust in all religions. Most centre on power and money.

13

Life on a Living Planet

WHAT WILL LIFE be like for you on a living planet, one who is no longer asleep but looking after the needs of her guests? What are your current worries, not enough water, is there global warming, are the seasons out of kilter? You may be worried about earthquakes, tsunamis and volcanoes, soil erosion, nuclear waste, and oil spills. These are quite different subjects but they are predominately about the Earth's outer surface, and each has a large effect on human lives. In one hundred years you have drained the easily reachable oil deposits, and a tiny number of people have made vast fortunes from doing this. They provided you with fuel to run cars that dirtied the air and began the process of global warming. They denied global warming existed and carried on in the same manner, looking for oil so you could burn it. The oil is a lubricant and part of the make-up of the Earth, which she uses to help her move and slide. It's easier to move her crust around if it is sitting on oil. You may say not all oil comes from earthquake regions, but you have to take a long view of this over time. Very few regions have never had an earthquake.

We spoke about the connection between actions and consequences, and your leadership is denying that actions which make so much money are doing any harm to people or the planet. Because this is humanity doing this, you can't shirk your responsibility. With the exception of Greenpeace and a few other organisations, you are complicit because you buy the oil, diamonds, and coal from open cast mining, fish from over fished seas, and allow other degrading practices. You use electricity from nuclear power stations, and you build cities in the path of volcanic lava in many parts of the world, or on the coastlines where tsunamis might strike. There are enormous populations sitting on top of fault lines world wide. Your actions have created global warming, and this unbalanced the seasons. Water is used as if it will never run out, and populations move to desert locations and plant golf courses. This is not sensible behaviour.

Of all the resources on this planet, the most abundant is water. You have more water than land, and these oceans generate rain through a pattern of heating and cooling. If you had to draw freshwater up out of the oceans in the quantities that rain clouds do, how would you do it? Instead you build desalination plants run on electricity, or you collect water when it rains on land. It all worked in the past and there was only regional drought, when global warming began to alter the heating and cooling patterns that produced the rain clouds, and began to melt the polar ice caps. You know this is happening, but you haven't changed the behaviour patterns that started the melting. The new Earth will have new energy, and the physical changes towards her health will follow later. Do not expect to wake up one day in December and find the polar ice caps miraculously restored. The energy will be there to restore them, but you will still be able to block it happening. This water problem is an opportunity to work for a solution across all international boundaries.

We have often said that the Earth can make it rain where it is needed, and crops are able to grow up out of the surface overnight. The seeds respond to her and everything she provides for them. If she were to clean up your messes, would you learn not to make them again? She will want to re-establish paradise for animals, plants and you, but you are responsible for your actions. There are currently clean-up projects in underwater reefs, and on the surface of the Earth. Humanity will need to be actively involved in cleaning up the planet. By doing this you will have changed the energy around your activities (back to energy again – it's the only thing that matters) and that will bring about the end of these problems more quickly. When the energy changes from dirtying to cleaning you will find support from unexpected places, from the Earth, plants, animals and elementals. We said in *Planet Earth Today* that the elementals had made a particular study of how to deal with nuclear waste. They are waiting for the opportunity to help you deal with it. Clean-ups are done step by step, place by place, and with no new messes being made.

Volcanoes, earthquakes and the resultant tsunamis are not a result of your activities. These have always happened as a natural part of the Earth's surface. For her an earthquake is little more than a large, rumbling snore. The years after 2012 will not hold any more earthquakes than those before. You are due to have some and they will occur. The big quake in California will be made worse by the fact that the oil has been removed from under her surface. Instead of a shock-absorbing layer of oil the land will crumble like pastry, and that is due to your own activities. Some large volcanoes are also due to release, and if you did not have so many people living close by it wouldn't have the potential to be such a catastrophe. These events have always been necessary for her, and you can't stop them happening. You can use your connection to her to sense when they will be coming and get out of the way.

Finally, tsunamis were less of a problem before the build-up along the coastlines with resorts. There were always deaths from them, but you now have many living right on the seaside all around the world. It puts humans in the pathway of the giant waves. Global warming has melted glaciers and raised sea levels to drown villages around the Indian Ocean, where people are at great risk of dying.

We write in a matter-of-fact way about people dying in natural catastrophes, but we are not indifferent to human deaths. You have the ability to listen to the Earth without scientific instruments and walk away from disaster, just like the animals do. You could watch them; they're listening to their Earth mother. Also we are aware that death itself is very gentle, the other side of life and not a stranger. No one needs to fear death, but we know the dead are sadly missed by the living. We love you, and we are sorry when you are sad.

The important relationship to remember in all of this is your interconnectedness to the planet and all life on her. She is hosting seven billion of you and will not destroy you or wipe you out. You are necessary to her as part of the whole web of life on the surface. She is willing to help you in your game here, and she does exactly that. If you were told the exact time and date of a natural catastrophe something crucial would have changed in the experience, because you elected to be blind-folded here. Time plays an important role in that blind-fold. Your experiences are approached from a position of unknowing, those are the rules. We are giving those of you who read these books background information or indications of possible futures. We know that it is the time for Earth to change and we are planting seeds of light and hope, so that light and hope will grow from them. Many will never read these but will act as if they had because you watered the seeds, and acted on the information. We look for ways to promote change towards the light, but stop short of making you act on our orders as the dark angels do. You are free to act as you wish with us.

Post-2012 will be a period of rapid change, a time when the increasing speed of events that began in April 2012 continues apace. This is life without obstacles, without everything you put in place to slow you down and hold you back. Humans are so good at hedging their bets and finding reasons not to commit 100% to their lives. You'll no longer be able to live successfully in that way. The new world will call for you to show your heart openly, so that if you divert from your truest path and highest good you will realise it, and others will be able to see it clearly. You will find wasting time intolerable, perhaps you do already, but everyone will feel the same way. You incarnated with a purpose and you will feel most comfortable when you can pursue it night and day. There will be relaxation and socialising, but you will be living with your attention focused on your goal.

If you consider this, you will realise that it is considered exceptional and single-minded to pursue a goal this way in the modern world. There are men (by far more men than women) who have succeeded in business by disregarding everything except their work. What if everyone was like that? You might think it would be quite unpleasant. Did you incarnate to be successful in business and make money? You incarnated to understand who you are in relation to other humans and God. You will need to focus on others as living aspects of yourself, with one soul and many, many bodies. Focusing on money will leave some in the slow lane while everyone else is discovering what it is to be human, and the material people will look very dull and one-dimensional. It also has the effect of skewing the flow of money to those who devote themselves exclusively to collecting it for themselves. Relaxing this effort will allow some of it to naturally flow back to others who are producing the goods that make the money.

That sounds like people will still be making goods and providing services. They will, and you will still want the specialisation of workers. You will support many different types of goods and services, from hi-tech to healers to farmers and fishermen. It will not be people just sitting and meditating all their waking hours, but interacting through working and living. People will not be trapped into a profession that does not give them pleasure, or outlived its interest for them. The ability to move to something new will have become easier and been accepted, and no one would say you couldn't leave a job that stopped making you happy. Even today you can leave a job that doesn't suit you, but people are afraid to move on. It's fear that holds you back now; that will have left you, and it is the only thing that stops you moving forward in the present.

How long can you expect this transition period to last? It will last a few years, but within ten years there will have been some serious shifts in society's expectations, and a restructuring of work-life balance. Following this period when new foundations are being laid, it will be easier to restructure the way people actually spend their time. It's changing the underlying beliefs that are the hardest. Changing the expectations sets everyone free to alter their lives, change their locations, their professions or their partners. It will be ok to change, or to remain. Happiness will be the measure of success for everything. It could be so today.

Let's take an example of a family where the woman works at a shop selling knives she made and designed herself. The children of the family are in school in the mornings, and after lunch are busy helping her and learning the trade of making knives. This gives them time to be together and get to know each other, and feel how much they are loved. There is a father who is out in the mornings helping on a neighbour's farm, but he returns to the house to cook and clean, and help in the shop. He brings home some food when it's the right season and preserves fresh fruit and vegetables for winter when the growing season has finished. He's back and forth from the home to the knife shop, and he spends time with his family and children. The children know their parents so much better than any today that barely see them, parents who are hard at work out of the house to make ends meet. There is enough money coming in from the shop, and enough food for all of them. When the children grow up they leave to find their own lives. One makes knives because she loves doing it and helps her mother in her shop. The other farms and makes leather goods on the side to sell. The skills learned from their parents about running a business and working with their hands has transferred to a new profession in the family. The key aspect is that people are working together and make a unit that is in contact with each other for hours everyday, and the parents teach the children. The schools are only teaching in the mornings because that is enough time for reading, writing, math and academic subjects. The life skills learned in the afternoons are taken out of the curriculum and taught by parents.

There are many parents today who are unable to work because there are no jobs, or they haven't been taught the skills to hold down a job. This deprives them of their reason for living: to have a purpose in their lifetimes. They are to be pitied for the waste of their lives, but not for the waste of an incarnation. Every incarnation has a purpose even if you can't work out what it is. Huge numbers of jobless people and their jobless children will create a flood of pressure in some directions. That is the point, to create change.

Your purpose as an incarnation of a soul of light is to be happy right now; you don't have to wait ten years, or ten days even. If you want to live post-2012 today, spend time weeding out the things that don't make you happy and moving forward towards everything that does. The more you let go of now, day by day, the easier it will be to help the entire soul learn the lessons of happiness. This is how you help yourselves move forward.

14

Global War on the Horizon

THERE ARE POSSIBLE futures, three of which we spoke about in detail in *Planet Earth Today*. The scenario of a vast number of people dying through war remains a strong possibility, as do the other possible outcomes following the war. The way these possibilities come about is through the individual actions you take today. Every action taken by a single person has a follow-on affect, so what you do today changes the entire world by tomorrow, although usually in a small way unless you are a world leader economically or politically. There is a huge nexus of timelines that have dove-tailed together in 2012, it looks as if every timeline meets in a single knot at the end of 2012, and then fans out again. You are not going to know exactly where you are on a timeline while you are incarnate. If you act in truth and light on one day a timeline towards light is progressed. If you keep acting on that pathway the line strengthens, and perhaps more people will join you on it. At the same time other timelines come to an end because they have atrophied. It's always like this, but there is not ordinarily a possibility for everything in one short year.

When we previously wrote about global war we indicated that many would die, and that the war would follow a period of famine and drought. The main point of the three possible versions of the war was to show you how the future is set up by the way the end of the war is brought about. One future was dark with never-ending tribal warfare and destitution on a dead planet, one was ended when the people realise where their actions have led them and join together to stop fighting, and the in final one the young soldiers themselves refuse to fight any longer. Each created separate, new futures some of which would be more pleasant to live through for humanity. Even the desultory ongoing war would teach you about the nature of God, but you would have taken the pathway that leads to misery, and the ripples that leave the planet would be those of sadness. To say that a game ends badly is to misunderstand the nature of God, and you would be implying that a part of Him is less good than another. We are not criticising God; our chosen role is to promote happiness and we remain true to the part of God that is light. It's our job.

You may consider your actions weak and ineffectual, but they are carrying so much weight this year that it's hugely important how you live, and what choices you make. The actions this year select the timelines for next year, or the possibilities of how the future years will turn out as everything begins again afresh. Although we wrote about a simple life in the small section above we could still see the future turn out differently. But we think that particular future is pretty sure to happen, and a good one to keep in your thoughts and help support that type of pleasant life coming to pass. This is how you deliberately create the future, by serious consideration of the life that would make you happy. Some young people already know they want to live near their parents, or in a town and not a city. Others choose the opposite and what we are trying to emphasize is the word "choose". It's a lot different from having others choose for you, or just drifting into something you don't enjoy. This is about understanding the power each person has to create their own happiness by being clear about what makes them happy. You will only be clear if you consciously sift through experiences and make decisions based on what you like. If you are happy, you are enjoying a successful life.

The scenario of vast numbers of young people incarnating just to die in battle seems horrific, but every life and every role that is chosen is perfect and leads to an advancement in the overall soul. The goal of the total war is to change the way people view life and the planet. Life isn't particularly precious at the moment, and many lives are taken so thoughtlessly. Fewer survivors mean more value placed on each life, and less use of the Earth's resources, the animals and the fish in the oceans. How big a shock does humanity need to change priorities, and help them accomplish the human soul's main goal? The object is to learn about yourselves as a small part of God, both light and dark.

If war doesn't come, and there are no guarantees, it is because a number of small steps are taken by individuals that begin to build a different future. What are the conditions that avoid global war? Love towards all your neighbours near and far, and reining in aggressive talk and tendencies. You may feel that people around you are talking in an angry way about others, and it is exactly then that words of love and reason should be spoken to diffuse thoughts of violence. Can you imagine a country's leadership going to war if no one backs them up, and if everyone is protesting? Wars are often ended when the people take to the streets to protest. Don't wait for military action to start but protest as soon as there is talk about it. If you can't get a group together go on your own, and keep going until others join you. If you are not one to take part in a demonstration, encourage and support those who are.

In the summer of 2011 there were serious riots in the UK, and a breakdown of order by young people who felt they were outside the general society. They were rattling the bars of the cage you feel you live safely inside, a place where there is order and the state looks after you. The state couldn't look after citizens for a variety of reasons, and that was a big point of the riots for us. It was another opportunity to change your views on how the state cares for everyone from the rioters to the property owners. Why should the state be in the position of looking after everyone? You are born with the ability to look after yourselves, but that ability is not being used by many people. Did you think that perhaps some of those young people were born to challenge the current system? There's a lot wrong with a system where so many people have no purpose to their lives. It's not a simple problem to unravel, but it is worth seeing that it is a waste that people have no chance to use the talents they were born with.

There are many people who just exist from day to day, eating, sleeping and passing the time because they have nothing else to do. This is true deprivation, to be deprived of the means to be happy. Your societies have institutionalised unhappiness, and it is possible to change if there is the will to change it. If the citizens desire it, why don't your governments change it? There is a problem with all governments in that they want the world to conform to their ideas, and not adjust their ideas to conform to the world.

One of the reasons why we see the possibility of a huge loss of life in war is the people your governments have marginalised. They add no value to the economy so they are seen as valueless, or even a burden on taxpayers. If you see them as part of yourselves then they are precious, after all you wouldn't harm one of your limbs if you could help it. We fear it may seem to be a solution to let many die to preserve the lifestyles and incomes of the few. You must be on your guard against this: if you think this is a good resolution you're being naive as you rip apart a single soul. Instead, show you value everyone's life.

War is an inferno of hate, and when you plunge into war you let loose without restraint hatred and violence. These are very dark energies and they build and build, and fuel the rise of more dark energy. The loss of life does not make life on Earth easier for everyone else. We recommend working together to find solutions based on love, and take into account the individual sparkle of everyone alive. When you look at a small baby, joyful and smiling, your hearts must break to see the sullen teenager that baby grew up to be. What was taken from that baby to create that unhappy young person? Hope was taken and the opportunity to live as part of the rest of the world, and not pushed to one side valueless. You wouldn't like it if it happened to you, and as you are all one being in many different bodies you cannot escape from this person's pain.

War is the absence of light. When you embrace the dark it doesn't matter who instigates it, or the reason; whether it is religious or political or economic. Those who lead you into war are not doing it because they are on a pathway of light. You can understand this if you think about it. They may be misled or misguided, but they are listening to the wrong whispers in their ears. You don't have to go along with your leaders who promote war, and the chances are they aren't telling you the real reason they want you to fight for them; usually money or personal power. You get to pay for it and then fight for them.

In the post-2012 years the split that we mentioned between those who are on a trajectory of light and those walking along below will help to demonstrate new ways of living. You have read the occasional story about people "downsizing" from the city to the country, and it's still unusual enough today to get an article about them in the paper. This is a lifestyle that is discouraged at the moment by so many tiny obstacles, mental, emotional and practical. Downsizing is going to become more common as people feel they are losing more than they have gained by their stressful lives in big cities. By walking away they show they do not value those lifestyles and it will lead others to question why they are still there. This will change the energy around the ingredients of a successful life, and money will lose importance to become part of the picture instead of the major part of everyone's life. The few with high salaries and vast wealth will continue their current patterns; do not expect real change there. If they weren't that fond of money they wouldn't work so hard for it to the exclusion of other parts of their lives. They are wrapping their hearts around cold pieces of cash.

The unhappy energy of discontent with a current standard of living can lift when more people swap to smaller lifestyles. There are a lot of people out there who have enough for all their needs, but never feel they have enough because they see others with so much more. They are unhappy, and yet they have all the ingredients for happiness already. This is another great disservice the wealthy have done, by helping others feel they are deprived of happiness by being deprived of fortunes. If all the money in the bank accounts of the wealthy were spread across the world evenly, it would barely improve the lives of those in the West. You are the wealthy ones compared to the poor of the third world. Their lives would be noticeably improved, and yours would not by an even share out of money. This is about letting go of discontent and fear of not having enough, and fear that you're missing out. Letting these go is all you need to have your spirits lift, and counting your blessings is a very effective way to change the way you feel. If you want the new societies to be different from the old you need to only to change yourself, and let that change ripple out.

The energy of change is enhanced by people living and adapting to change, and even by writing down in books like this one the effects of change. Once it is in print it is "out there", and that begins a process all by itself. It is one reason why we write books and put them into print, especially prior to putting them into electronic media. It's about making it real and initiating a change in energy.

Change arrives and people block it from their lives through fear. Going back to fear being the epitome of darkness, you are blocking light with dark. Is that what you want to do? Since light is joy, change and flow will allow you to experience joy more easily. There is a pent-up dam full of change waiting to take place in your lives, waiting for you to unblock and let it in. Some are more ready for this than others. What helps some people is to just start making changes, in your daily routine, or in your weekends; shake things up a bit. Change the furniture around, go away on a break. When you come back after a week away you are already doing a few things differently, it doesn't take long to change the small things successfully and start a process of flow. Find a way to wriggle out of the grip that fear of change has on you. The biggest help is to get rid of old attitudes, possessions, fears, etc.; they are the small blocks that keep you in one place. Cleaning out your garage or shed is as good a place to start as any, it begins a process of flow towards the light. You will be learning how to live with flow on a planet shining with light.

We talk a lot about change because it is such a strong energy and the winds of change are what will blow you from the old Earth to the new one. Can you conceive of how hard this wind will be blowing? You will be ripped in half if you can't run along with the wind, because your feet are buried in mud. 2012 is the year when these winds start to blow, and pick up velocity throughout the year. In the wake of these winds will come turmoil. You have seen the effects of strong winds, when tree branches are broken off and rubbish and leaves collect in corners. The effects of those winds are mild in comparison to what you will find in your lives in 2013 and the following years. You can't reattach tree branches to trees, but you can spend time cleaning up. Which of your branches need to be pruned before that happens? Do you want to be moving forward with your lives or cleaning up the past?

Some people will shed their burdens automatically and find they moved right along and never think about it at all, and other will have had to work harder during 2012, depending on the individual. If 2013 finds you with broken branches all around you, then just like a tree you will carry on living. We're pointing this out now to help you prepare so the damage is less, and you are free to enjoy your new life and the new planet. A few people will be too rigid to weather the storm, and instead of bending with the prevailing wind, they will break.

2013 through 2018 or thereabouts will be the clean-up years, and there will be changes taking place to cope with the new planet. There will be a huge number of energetic corrections taking place that feed into new physical realities; you have the chance to start moving in a new direction right now. You will strengthen your personal path to the light, and strengthen and help create strong timelines of light. This is what we are looking for from light workers, as many people will be confused and unsure of what they should be doing in their personal lives. Your ability to hold light will put you in step with the planet, and others will be drawn to you as a way of being helped with their own process of change. You will lead the way by the energy you hold, and by the happiness you exhibit. The Earth will have partners in joy.

Part Four
Days of Light to Come

15

Earth Rejoins the Universe

THERE WILL BE days of turmoil and change after 2012, but one day many changes will be completed and humanity will move forward together. This depends on the small day to day actions of many people, and their resistance to actions that are not based on love. This section is to paint a picture of a world of happiness, not a world twisted into self-enrichment or continual warfare, and the final stage of your game here will be to reassert the genuine love and kindness that exists in the human heart. There are far more loving people on Earth right now than those with dark hearts. Whilst you may have stepped back and turned away from many who frighten you, there are enough good people to change the world by working together. The minority will continue to harm others, but there will be less tolerance for letting them get away with it.

Let's start near the beginning, around the years 2018 to 2020. There will be nations at different levels of change, and people will have turned against the institutionalised practices that trap people into uselessness. Eyes will have opened to how damaging it is to isolate others out of the way. Barriers will come down, and these unvalued people will be drawn back into the world. Their energy and contributions will begin to be felt more widely. The current energy in the big welfare housing estates is stagnant as these people have no role to play. When they are drawn into society their energy will become unblocked and start to flow. What did we say about flow versus stagnation? Humanity does not want large groups of people with no purpose to their lives. The increase in the flow of energy by including these people will help everyone match the flow of the New Earth's energy, and the divisions between you will become less.

The ultimate goal is to remember that you are all one. To do this you will need to look into the eyes of others and see yourself reflected back. It's one of Archangel Ariel's techniques for enlightenment from *Planet Earth Today*, and well worth practising every day. When more and more of you understand that you are one soul you will change the way you treat fragments of yourself. Treating others the way you would like to be treated yourself is the "Golden Rule", and if you practise this rule you will break down barriers between you. Before you ascend you are going to have to remember this. When you care for every person on Earth with the same amount of love and concern you show yourself you will reach personal ascension, and that will add a sparkle of light to the overall soul. Those who ascend in advance of the rest hold the energy of light for humanity, whether they choose to incarnate again on Earth or not. There are twenty-eight human incarnate ascended masters on Earth at this time.

Your original goal was to find God from a position of complete unknowing. You would have to find within yourselves the spark of God that gives you and everyone else life and existence. This spark is in everything, and you are connected by sharing in common your origins in God. Remembering that you are one with each other is the penultimate stage of the game; remembering you are one with all of creation is the final step. When humanity reaches that point together your soul group and the other souls here will join the Earth in ascension.

Your best chance of reaching ascension is with the help of the New Earth and the new energy she brings. Without hindrance from the angels of darkness you have only your own natures to contend with. If God is everything, then darkness has come from Him also, and you reflect the light and dark present in the Creator's nature. We are not suggesting you give way to your darker natures and continue along the pathway to separation. We want you to enjoy your lives and draw together, while caring for each other and the planet. There is support for you to move towards the light from the angels of light, from the planet, and from plants and animals. There is support from the human ascended masters who maintain a core of light in the greater soul. The human soul will find it easier to make progress now, and can move ahead more rapidly.

How will plants and animals change? Put your hand on a tree and tune into its emotions. They are so happy now, and bursting with joy. If you live near trees you will be influenced by the joy surrounding you, and this is another reason why we are not fond of cities. There aren't enough trees in them to make a difference. The animals feel the Earth through their feet and are basking in the light rising from her surface. They're anticipating the changeover with great pleasure because she will be healthier afterwards. They're helping with preparations by contributing their own energy of joy. This is a happy, happy world again and whether you can feel it or not, it is changing for the better.

What can you do to help? Everywhere needs Earth healing groups to burn away the darkness and help her by channelling light into her. Those who participate leave with their energy restored by the activities, and they have a good time so their spirits are high. This is what we mean by the little actions that take place in a life. This is a good example of a small amount of time having a big effect. The Earth knows that it is humanity taking part in those circles and that they care enough about her to do so. These groups will be needed on the New Earth for healing and communication.

Another useful activity is to spend more time outside! With your feet on the grass and the sun shining, or in the shade of a tree you have a chance to blend energetically with other species and learn from them. If you walk down a shaded lane your energy fields pass through those of each tree in turn. You exchange with them information about yourselves, and you are less insular. By staying indoors you miss all this, and your progress is slow. It's yet another reason to be sorry that many children no longer play outside, as they are deprived of so much wisdom. You have to start learning somewhere and learning from other forms of life on your planet is a good place to start. The information you learn outside is about the wider world and the universe, and how you fit into it.

Universal information is pouring through Avebury Stone Circle in the form of symbols, or codes. There's a lot of catching up to do with a universe full of old friends and neighbours. All the insights they have learned about themselves will be downloading into your planet, and this will accelerate some of you to ever greater self-knowledge as you find a key piece of information circulating in the Earth's energy fields. Some of the more technically minded will absorb new ideas for inventions that will seem dramatic and unbelievably creative. A future full of space-age inventions will have finally arrived. (How can you say someone owns these ideas that came from another planet?)

Technology is something that humans rely on, and there are many who hope that it will rescue them from your current problems. Oil gone? You won't need oil, because we've developed a new substitute for you to use in your cars and houses! Or so you hope someone will say. This isn't a time to be imagining more of the same, an oil substitute to burn or expensively manufacture. This is the time to allow ideas to form with no preconceptions; as you will find there are a number of new ways to get from A to B when the new technology begins to percolate through. You have made advances in communications through your computers, and can now freely video-phone your contacts through Skype. Once you could barely imagine making such a phone call. These are the types of leaps that you have to be ready for. We quite like the idea of teleporting people from place to place, but you'll have to wait a little longer for that. You will have to have altered your physical structure before you can be "beamed" anywhere like they do in Star Trek.

Another import will be the resumption of the universal rule of law. You have laws here, passed by your governments, and some activities are "against the law". This is not the same as justice. Laws were not developed universally to protect private property from others, or to hide behind. Justice is about right and wrong, black and white. The universal court examines plaintiff and accused with the same intensity, and justice is delivered. Sometimes the full penalty of law is applied against one of the parties to keep other people safe. They don't use the death penalty, but the remainder of an incarnation may be energetically spent inside a clear box. It is not lawful to kill or attack others, and you know this. You can be very slow to intervene in another country where the government is killing its citizens without penalty, as in Syria right now. When you do not intervene you are complicit in the crime.

The universal court is available to all. If you wish to address the court you may say "I wish to take this person or entity before the universal court for attacking me and I ask for justice." Angels of Justice run the courts and you and the incident will both be examined inside and out for the truth. Your higher selves are the ones who are present as well as independent witnesses. When it has been reviewed a judgement is decided on and administered. Because this court has not been accessible to you for eons your own justice system has been twisted out of recognition. How else could you say you have no "right" to intervene in a killing situation?

Another benefit will be to return you energetically to your place as a unit in a larger galaxy. The Earth has been bobbing about without an anchor, being subject to every breeze. Her rightful place is connected into a web of relationships. When she is part of the larger web her stability is greater, she moves in concert with others instead of by herself. This is important! Here is one of the energetic changes that will make a difference. Increased stability for the Earth makes her job easier and gives her far greater support in her life. You will be living on a strengthened planet and that will feed through to the type of actions that are supported on her surface. Would you be more stable if you stood on a rolling ball or if you stood on a plank? This is going to matter greatly to all of you in the years to come. Sometimes part of the reason your lives have been so unbalanced is that you were living on a disconnected planet. That's all changing now as the year's progress.

In many ways life here is primitive because you have chosen to play in a splintered form. The messages arriving now are telling you that everything in the universe is one, and together they make up a whole unit. You may not be broadcasting this yourself from Earth because you have blanked it out, but soon you will be receiving this message non-stop. This is the one we attach the greatest importance to and that will give you the largest push forward. It's the message that will cause you to say "Enough! Stop that now!" We're looking forward to that day, and it's not that far away.

The new technologies will be planet-friendly, and revolve around helping humanity rather than advance warfare. The universe is fairly old now, and what you will be exporting to other worlds in the line of armaments will be of little interest to them. The most useful information from here will be the vibrational flower and plant remedies made from your native plants. They will be appreciated by those living on other planets. You developed them here in isolation and they have an Earth-flavour to them. Some planets will be able to duplicate this technology using their own plants, but others will not have plants growing on their surfaces. There are no green plants on gas giants like Jupiter.

One set of information that will be of interest from here will be broadcasts by the crystals. They have been keeping records of their own experiences and observations and these will be received early on by everyone because of the clarity of the crystal's signal. They will be telling the story of the soul group who blind-folded themselves to find God. The other planets are very interested in this particular game, as are the soul groups living on them. Resuming broadcasts will mean something is changing here, and the darkness is lifting. This is a large enough event to trigger the beginning of the end of all the games on all the planets. The movement to draw together will begin, and that is how the universe comes to an end one day.

If you have any doubt about the darkness lifting on Earth, listen to your news broadcasts. There are revelations and secrets being exposed, and that is changing the way people behave towards each other. You have already made some changes that are reversing the swing of the pendulum. That makes it easier for you to keep in step with the new energy. There are some very big, core secrets to dig up and expose, and once that process starts you will have a steadier forward motion. We are referring to the people who run the governments from behind the scenes for their own financial profits. The pot of money paid in by taxpayers has proved to be too great a temptation for some, and it'll all come out in the open. Don't hide your eyes when it does but keep pressing forward until it is all cleaned out. Remember you are lancing the boil.

Earth has been in isolation for thousands of years. You have had your ebbs and flows of civilisations; you were highly advanced in Atlantis, and less so as medieval peasants. You have been trying out different experiences. When one set of experiences led to a dead end it collapsed and you had a period of adjustment. The fall of the Roman Empire is an example, it ran its course and then you had a number of years of trying smaller societies. So what is going to happen this time? It won't be necessary to repeat the Dark Ages, as you already lived through that. Remember there is a greater human soul, only part of which is currently in human form. There is a consensus when you are dead and plans are made that include the entire soul group which you forget when you are alive. You are heading into something new that humanity has not tried before, and it has the potential to be your final stage.

Humanity has been locked into a long period of inequality, where which sex you are, the colour of your skin, or where you were born kept many from sharing equally in society. You've become used to this and many of you have given up on anything improving in your lifetimes. You will realise the extent of the inequality when you find out where all the money is being hoarded. If there existed a big pile of money and you stopped by to pick up your share in a basket, and just took your share, there are a few others who would try to take all of it home and dare you to complain. You have been harbouring these people in your midst and allowed them to take far more than their share home and lock it away. This is the love of money, and the greed that has been attached to money. In the last fifteen to twenty years the greed has grown exponentially while the world's wealth has been siphoned into sterile bank accounts.

Money is a medium of exchange. If you buy shoes, the person who sold them to you can take that money and buy food. The grocer buys some flowers and some books, and the book seller goes to a restaurant. It's the same money being passed from hand to hand until it is locked into the bank accounts of someone who has more money than they can spend. Then it stops circulating and the rest of you are poorer. The very wealthy are taking everyone's money and skewing the world's economies. They have become a burden almost too heavy to carry, and you are all working to support them.

When we talk about the energy changing and the lack of support for something, we see this affecting your current economies and the money flowing to just a few families. You are going to face upheaval in the distribution of money, and what is considered acceptable wages for goods and services. Allowing money to go back to its original purpose as a medium of exchange will free it of the weight of greed. This renewed flow of money will allow so many people to breathe freely again. You have lived as economic slaves, and you have not been able to enjoy the fruit of your labours. Equality is a precondition for uniting as one; you can't treat another as an unhappy slave if you know you are all one. Taking advantage of other people and laws of property keeps you separate from one another.

16

One Soul, Seven Billion Bodies

IF YOU HAVE A decade of war in your future it will result in lower population levels. These lower levels will be permanent and will reach a point of equilibrium where there is enough space for humans and enough space for all the other species. You will have found balance in yourselves which will make you comfortable with fewer babies. Part of the reason for the large families you now have is this lack of balance, and feeling that babies will provide something you are missing. They are very good at providing love and joy, and we foresee the time when you will have enough self-love and joy in living to stand alone. You will not need numerous children to love you and will have smaller families where each child receives a lot of attention and care from their parents. This happens today in some families, but in your very large families the oldest children are raising the youngest while the parents are working frantically to provide. The need for this large population will be gone. This will happen over a number of years.

The largest adjustment to population levels will be civilians and soldiers who may be killed in global war. We do not wish to see this and hope you will find a way of avoiding it, as your population can decline naturally without war. Your balance as individuals creates balance around you, and brings about calmer times. One of the main reasons to seek out joy and light is to avoid carnage where you risk so much energetically. Death brings sorrow and the waste of young lives, and war doesn't make you happy. We see a very strong possibility of war in the not-too-distant future. War is likely to be started by those who are trying to maintain control of the world's economies, and there is a lot of money to be made in armaments. These people are not energetically aware and won't understand that the Earth is now radiating light, and that building a dark empire on light is almost impossible. These are constructed in the shadows, not the light.

Smaller populations with fewer people will provide a greater depth of soul resident in each body on the planet. Your greater human soul is divided into seven billion bodies at the moment. The original number of humans on the planet was far fewer, less than one million. Each of these experienced life, then died and came back for a new life. They learned by living, and kept their soul separate from the rest, and they learned all their own lessons. This is the main reason your game has taken so long, as you don't pool your experiences in the same way as other soul groups. Some of you have learned enough to become ascended masters and without the need to incarnate. Those souls guide the living and wish to help humanity ascend. The original souls present in the beginning have made the choice to divide and live in more than one body. You have less soul operating more bodies than ever before and you can see this if you look at the flicker of soul attached to each body. Some have a great deal of soul in this lifetime, and others are making do with much less. To us this means some of you are struggling on a spiritual level as you have so little spirit. When we said some people would never understand what happened in 2012, these are the ones that will not get it. Before anyone thinks "I have more soul and am therefore better than those people," we would like to remind you that would be like saying one leg is better than the other.

The problem with spreading one group soul into seven billion pieces is that it is much diluted and alters the behaviour of people. It definitely makes them behave with less humanity. This will all balance out and come right in the end, and the shrinking of families will come as a result of balance, not by preaching at them.

What is the effect on the planet, on the light and dark balance and the human soul when churches promote these large families? There never was a need to deliberately enlarge your families; they were fine as they were. This could only take place against a background of greed, where everything on the planet belonged to humans and is there for the taking. Enlarging families has been part of the means of weakening them by rationing out love and care among so many. There are selfish motives of influence, power and wealth behind these policies, but it takes a strong individual to move away from the church that has controlled them all their lives.

What do religions teach you? Many traditional religions teach you that God is out of reach and that you are not worthy of direct contact with Him. They say that you will never be good enough, so a super-natural intervention will be necessary in the form of saints, gods, ascended masters, angels and others. This introduces a hierarchy of beings who deserve to be closer to God than you are. It would be like one of your toes being closer to you than the other toes. It's just not true! If you are a tiny part of God, you are all the same distance from Him with the same connection to him, just like your toes. We object strongly to religions because they are presenting a set of rules to reach God, and they are not based on truth. They are based on lies, and we hope you understand by now why you want to avoid lies. The advantages of religion belong to those few who are in charge, telling the others how to live and collecting the money. If it's not about love, if it condones power of one sex over another, or violence, then it has nothing of truth to offer you. It will not lead you closer to God.

A final thought, if you are all God, then of course it is true that God is on your side in every war. He will be on everyone's side in a war, and it would be helpful if you ignored these statements whenever you hear them. "God is on our side" has been trotted out for every war for centuries as a justification to get you to fight.

What if you had never developed any religions at all on Earth? When you chose to go to war for gain, or for economic reasons, you may have understood what you were fighting for. Northern Ireland is a good example of an economic war, where one religion was the native population, and the incomers were of a different religion. Jobs and wealth were kept by the winners in this war, and the two peoples fought for these. They weren't fighting over how Jesus was worshipped; they were fighting along ethnic lines for control of the local wealth. This dispute is mirrored all around the world today in religion after religion. It is a blanket of darkness, and it has kept the light from reaching the planet. Religion will try to re-establish itself more vigorously than any of the other forms of darkness. They are enormous bastions of control and power, and they use fear to keep hold of their followers. An emotion like fear provides them with greater strength than money.

The planet is now producing its own light, and we are looking forward to this light burning into the religious strongholds. There will be enough light rising up to expose them and reduce their credibility. We know that when this layer is burnt off, with fewer rules to follow, and with less guilt over failure to meet all the requirements, people will be happier. Religions have been a method of controlling the masses in each country while the earthly benefits have gone to the leaders, and the energetic benefits have fuelled the number of dark angels on Earth. If you stopped using those stone buildings to broadcast worry and fear across the land your countries would be happier places. In the past few thousand years there have been many new religions trying to teach you how to find God. You could not look in a less helpful place, for your Creator is inside you, living in your heart.

A final word on religions: there are a lot of good people sitting in churches, synagogues, temples, and mosques every week looking to be a better and happier person who is pleasing their God. Maybe you are one of them. We know that people have spirits and are spiritual, and they want to know more about themselves and their Creator. Their parents brought them up to believe that what their religion tells them is true and that it is the only pathway to God. We love people, and we see their thirst for knowledge and the desire to be spiritual beings. They need more than food and sleep to live. What we are saying is that those in charge of these religions are not the same as you; they are too often in it to exercise personal power. You give them your power, and they are happy. We would rather see you happy.

17

Applying Light

NATIONAL GOVERNMENTS are another area of impenetrable darkness; all of them are putting on a show on the surface, which is the tip of the iceberg. There are acts taken in your name that you would never agree to, research funded and money handed to the friends of the political party in power. This is how you are governed, and once one party arrives in power they exercise it through patronage and giving out salaried jobs and contracts to their fellow party members (of course to their friends, never those who opposed them!) The changeover of staff is enormous and makes the stakes very high for those who win elections. People will owe them favours for the rest of their lives. These same men eventually go on to the boards of most big businesses and run the economies of the countries, and cross back and forth into government over the years. The people who run your country politically and economically are a single small group of men and a few women who sit on numerous boards and back-room committees.

There are departments that stand-alone with minimal supervision inside of government and one of the obvious ones in the US is the CIA, run according to their own purposes. There again a few men operate an organisation that alters millions of people's lives around the globe, for reasons they have decided are best. If you are unable to find out about these secret organisations you fund, how can you judge whether they do more harm than good? The CIA is closed to you all, but its energy is like a vast black spider that spins it webs. It has a controlling foot in many places, and no outsider knows what it is up to. The same goes for many of your military hierarchies. They can often be working for themselves and their personal prestige and not their countries. What good is an army that never fights? It loses power. Until three or four hundred years ago governments were so frightened of a professional standing army that they would not allow one to exist in peacetime. They had the example of Rome to look back on where every Caesar was put in place by the military. Have your leaders been put in place by the military? In many countries yes, but even in the West would you ever know?

Recently it has not been the military that put rulers into place in the West, but the fear of newspapers and bad reports. You have only yourselves to blame for buying these papers full of non-news and unpleasant gossip. The Leveson inquiry in the UK is slowly reviewing the shortcomings of the press and their abuse of power. The press has been feeding the most unpleasant type of stories to the population for generations. The public's appetite for shocking news has been insatiable and addictive. These papers have been like drug dealers pushing your latest fix. Supporting these papers, magazines and TV networks with your money has not been good for you, or an activity of light. You pulled a blanket of darkness over yourselves. There was a time when newspapers and journalists worked to throw light on a story, the most effective and famous case being the Watergate scandal. It rightly exposed the lengths President Nixon and his party would use to stay in power. Those journalists and the Washington Post took a risk in digging into that story, but without them nothing would have been known. This type of journalism has shrunk almost to the point of non-existence while the mind-clogging trivia of celebrities' lives has taken its place. In Rome the masses were diverted from their lives with circuses, you are diverted by celebrity gossip.

It numbs you and keeps your attention focused elsewhere.

Having said that, a lot has changed with regards to the press and there are many coming forward to expose how they abuse their power. This is how shining a light into a dark subject works, people begin to find out what has really been happening. They will have the chance to choose for themselves whether they want to keep buying one of these newspapers or not. The news media has been exposed as manipulating the police and governments, and those organisations were put before the rights of ordinary people. When they want to resume manipulation they will know there is already a light shining in that place, and they will have to look harder for a dark corner, and it may not be as dark or easy to hide. It's harder to deceive people in broad daylight, and now there are those who know where to look. It is illegal to print lies about people, or say them in public. The Leveson inquiry is investigating the invasions of privacy through accessing private phone messages by hacking. These are abusive practices and the theft of a person's privacy and reputation.

The change in the relationship of the press to government is one of the biggest stories of 2012. In the future, government ministers will think harder about openly handing so much power to the media. There's a vast difference between responsible reporting of events and manipulation of stories and emotions. The press and governments use each other, and the political parties know that if a newspaper with a large circulation comes out in support of them hundreds of thousands of votes across the land may be delivered. The trade off is handing power to the newspaper owners who can exercise a controlling interest in government without standing for election.

What's hiding in the dark that will come out in the light of change? As each dark secret is exposed to a searchlight, others will throw up barriers and bluster, challenging the public's right to know. Because it's your money you have the right to know how it is spent. We have talked about light in the abstract, but this section of the book is about how light is being applied right now.

The Leveson inquiry is a good example because by exposing the influence of the press there will be a genuine change in the way you are governed in the UK. Some who have exercised power without being elected will now be excluded. This is how truth can be applied in your lives and is something you all can use. This is how the new flow of energy will help uncover lies and put right those things that are very wrong. It's no time to put your heads in the sand and let others do the hard work.

18

A World of Slavery

MANY OF YOU have been very careful what you say about those you believe have power over you; the police, wealthy people with access to courts and many others with their privileges. You have been cautious to keep yourselves out of trouble, and failed to challenge those who were exploiting you. It's been very easy for those at the top of the pyramid to draw the wealth and life from the lower economic levels. Sometimes there is a change-around and the bottom becomes the top and it behaves exactly the same, or even worse, than the ones who were there before. When did it begin, when did you feel safer by giving your power away? What would happen if you stopped handing your power to someone else?

Fear and unhappiness are the fuel for the dark angels, the more that is created the stronger they become, and the more they are able to shape a planet to support their colonisation. Humanity's game of blind-man's bluff created an opening for them to come in and use your soul group as food. They began to multiply and grow stronger. Many of the strongest sit in the places of power in your world, and as your short lives go by these entities remain. They move from being to being, looking for someone with little integrity and use them to influence the actions of the government, religion or corporation. We feel that you know already who these particular people may have been in the last ten years. They say one thing and do the opposite, usually while smiling. What is your response when this happens? The majority of you let it go unchallenged. We can think of instances when the truth is pointed out to you in news reports, and the next day it is ignored.

Slavery is when one group of people controls another group and forces them to work, but the slaves don't benefit from the value of their labour. The profits from the value of the goods and services go to the owners. You are exhibiting all the attributes of slaves, not of free citizens. You are frightened of what will be done to you if you stick your neck out and you hope someone else will do it. Not since Atlantis have we seen such slavery. Perhaps you feel you are free, and we are exaggerating. We would like you to consider serfs who are tied to the land working for someone else's profit, and slaves whose hard work brings them no share of the rewards. Many of you are having trouble making ends meet today even with two wage earners. Your only hope seems to be to climb the same ladder that others have, and step on the same faces on the way up. Once you have made it into the ranks of the well-paid then you feel safe.

Those of you who are earning good salaries, do you feel safe challenging those higher up the ladder than you? Do you worry that you could lose your job and money and end up back down at the bottom? We are talking here about separation, about the other human beings that do not count, that you can't imagine as another version of your soul in a human body. If you continue to separate from each other you will not draw together, and this is why you are here. By separating you feed those who would enslave you in unhappiness, and by joining together you draw towards happiness. If you consent to continue exactly as you are, then your game ends in perpetual misery. You will still find God, because all paths lead to your Creator, but you choose the pathway of despair.

We are love, and we are joy, and in our opinion that is a good way to be. We come here to teach you because we empathise with each of you and want the best for you. The light portion of the universe reaches out to you and wants you to return to them in happiness, and the dark part wants you to realise that you are part of God, but through experiencing pain. Throughout all of this game you have freedom of choice of action, and you learn by day to day living. It's your game and it will end one day based on the actions you take. Our wish is to see it end in joy and have you join us in light.

In the coming years you will have interesting times as one exposure feeds into another. Please can you each keep up the pressure, and take it as an individual responsibility to not turn away when certain areas try to stay hidden. It will happen faster and more thoroughly if there is a groundswell of people pressing for the truth to emerge. Ask questions, write letters, sign petitions, talk to your friends and remind them of unanswered questions so they will not forget. It's a responsibility of being alive, of being a light worker that you all share. It's too big a job to leave to a few working alone. They are better as the spearhead of a large supportive movement. You can help as much or as little as you wish, as long as you don't walk away and leave it to others to do for you. You incarnated to be alive right now while all this is happening, and you all have roles to play in clearing out darkness and bringing in the light.

We expect to see the Houses of Parliament, the United State Congress and other major government bodies exposed as a haven for those that are there just for the money. There's a lot of money sloshing around governments, and there are a lot of special interests that are looking to receive some. The easiest way to receive government money is to influence legislation to work in their interest, and for that they need someone to propose the bill for them. Gone are the days when you had public servants, you have reverted to a world where you are taxed like peasants and you don't know where the money is spent. There are a lot of good uses for tax money, but there are also wasteful, irrelevant uses. Before we sound like a political party, we would like to say that there is so much corruption in the way you have been governed for years that no party can claim to be more pure than another.

In the past it was not considered wrong to ask someone to work in exchange for money. If you look back to the 1930's under Franklin D. Roosevelt in the USA he paid people to build dams and other projects for the public benefit. By the time World War II began the country was pulling out of recession. After the war it became politically incorrect to ask anyone to work for the money they received from the government if they had no jobs. This snapped another vital thread of life, the responsibility to stand alone unsupported. There is no way these people can be balanced in their individual lives, and live as rounded people using their talents and living out their purpose. Having no purpose robs them of this lifetime, and it robs you of a balanced society.

Your money paid in taxes could be spent on many useful and worthwhile projects that employ workers, but the way these projects are designed they hide where money really goes. The money finds its way back to the heads of corporations in their large salaries, for contracts they have received from the government. You have no one to stand up for you on the inside and expose what is happening, and it will have to come from outside bodies like journalists. The beginning of the clean-up process has started with journalism itself so that you will feel you can trust them again and believe their stories.

You have many governments in existence where it looks like the wolf is in charge of the sheep, and in the case of Syria that wolf is ravaging the sheep. There is a long history to any country and its problems, but if it were your leadership killing off the population would you sit quietly and let it happen? This is an extreme example, but at the time of writing it has been going on for over a year and no other country has stepped in to help stop it. Violence against a section of humanity hurts everyone alive, and you cannot separate yourself from the results of killing. There is too much pain created, and it could be stopped or prevented. This kind of savage killing of a country's citizens is one of the worst scenarios, as a government should serve the people, not themselves.

What kind of person kills their subjects; a person with a warm heart full of love? What kind of person are you, and do you feel that you can help someone else even though they live on the far side of the world? Who will help if you don't? This isn't just about Syria; this is about everyone who is mistreated or abused. If you stand by and do not act you are complicit. One burst of activity in the next few years by many, many people will change everything. It will seem easy once you begin to flow with the prevailing energy.

By public demand there will be changes in the nature of government, which will go as far as public oversight of government actions. You do not have public oversight at the moment. Military plans are not made public; they are in place to maintain the power of career officers and arms manufacturers. When offered a war the public usually says no, and they must be manipulated into agreeing to fight. There are other areas that have ballooned out of proportion to what is necessary, and they can be reduced. Whole research departments exist that would make you shudder if you knew about their activities. There is also an atmosphere of "not my party's fault" in politics, where the blame is continually laid at someone else's door. This is going to change and they will be held responsible for their actions, and this one change will lead to many others. They will have to put right what they destroyed in the name of ideology and money. Who will hold them to account and enforce responsibility? People like you.

We look to see nations where governments are more transparent and the money is spent on communally organised public services, like schools, roads and hospitals. By hiving off superfluous expenses they will have to let go of their power to award unnecessary contracts, and raise taxes for special interests. The swollen government departments will have returned to workable size and it will be easier to see exactly what they are doing. Because they will be open to view, they'll have to be truthful about their activities. That will be the moment when the lies stop and the truth begins, and the walls of society will cease to be made of lies. It is possible, it will happen, but you need to keep pressing for the truth.

19

Who Are the Light Workers Today?

YOU WHO ARE light workers look like candle flames to us, and you come in different sizes. Some of you are very big because you are using your light; and you need to use it to see it grow. You chose the role of light bearer and you have a clear feel for the truth as it is the same light that you hold inside yourselves. You vibrate with light and when you come up against lies you know you're being lied to. Others do not carry the same light you do, and they are confused by lies. If no one points out the truth to them, they believe the lies and get more and more ensnared in an unhappy web. Those people have their own roles to play, but if you drift and never use your reason for being here, you won't be happy because you will be unfulfilled.

Light workers are people who have lived through a number of incarnations and learned by experience. They learned the importance of action, and the importance of speaking out. Sometimes this got them killed, and they are carrying with them today the fear of danger that action and speaking out brought in their past lives. It is not unreasonable that it holds you back in this lifetime. You died back then, but here you are again today holding all that experience. The main thing you learned was that hiding and drifting wasn't really living.

2012 is very different from all previous years. Where you were struggling to be one point of light in a dark situation in the past, there are a host of light workers here now. You got together in one of the resting times between lives and arranged a mass incarnation of all experienced souls of light, and other's stood aside and ushered you through. You are here in various age groups, but the greatest numbers were born after 1986. Today if you stand up and point out lies there are a planet full of people to support you. You are not going to be lynched by an angry mob, as everything has changed to support the light. This is a reversal of the energy of the previous millennia, and you are safe to be the person you incarnated to be.

All light workers had a hard time somewhere in their past, and when we talk about letting go of what no longer serves you we include these old experiences and beliefs. Times have changed, and you don't need to live through the unpleasantness again. That experience will not be repeated now; you are finally here at a time when your inner light shines and you can let the fear of action go. How do you release a past-life fear? By looking at it and acknowledging its existence and letting it drop away. If you say it and mean it, "I release and heal obstacles I have put in my way of moving forward" is a sentence that can help.

Humanity is here playing a game, and you participate in the game by living your lives. There are many reasons why people hold themselves back from living, but it's time to let go and enjoy the experience. This is your contribution to what will be taking place here in the next few years. We are talking about human-only society; you wouldn't expect anyone else to come in and sort it out for you, would you? This is how you play the game of life. If you let go of your fear you will be having a great time living through this period of intense change. That's the main reason why you are here.

There are people who don't have flames in their hearts, and there are many here now that have no idea about light and dark. They can't tell the difference between them at all and will be of little help to you. They will be helped the most by creating good, solid areas of light where you maintain the forward flow that leads to joy. You create this in your own energy fields and person, and by joining with others who are the same as you there is a bigger area. Who are your friends of light, and can you plan something that makes you all happy? It's that kind of activity that will increase the momentum towards happiness.

One last thing before moving on, those of you who have a clear goal, such as "I am going to start bringing the light into my own life in every way possible;" will carry more weight than those who drift. Drifting gives up the power you were born with and has very little energy or effect on others. Sinking into flow and living in trust, meeting new opportunities and people and interacting with them is not vacantly drifting. It's crucial not to confuse the two. Someone who is drifting may bump into opportunities every day and not stretch out a hand to take them up, while others focus on what has been brought to them. It's about intention, focus and activity. If you are reading this book you did not come to drift, no one has.

20

An Army of Light Workers

YOU MAY THINK that the Western democracies are the "good" nations that have the least to correct, but you would be wrong. The greater the wealth, the greater the abuse of power has become. Turmoil and change will take place in your personal, economic and political lives. It is not altogether obvious to you right now which country has the farthest to fall, or has the greatest changes to live through. Chances are you think it is not your own country, but a foreign one with a different type of government. You will live through a great period of change, and it will be personal to you, and the only reforms you can influence will be those in the country where you live. It will be best to focus on what you can do, rather than what others should be doing.

You will work to open up your own nations to light in the beginning, and other countries may be holding onto practices you already put behind you. Twenty to thirty years in the future there will still be nations that are struggling to throw off darkness, and where people have set themselves against their own governments to bring about change. This is when you may see fighting between nations start, and when you may have to resist these other countries or your own government. Again, we need each of you to choose to play your part in your own lives. Stand together and don't allow anyone to fool you. Building up your country in light is the most important thing you can do in the coming years, and it will have a greater effect on those other nations than you can know. It's about balance and the sheer quantities of light present on Earth at the time.

War is present in the human race because you don't believe you are all one. That means it remains a possibility right up to the time of ascension, and it will only be avoided by the actions of peace-makers. We have witnessed private individuals, and how often it is mothers, who have achieved peace in their countries by concerted action. These are your role models and the people who deserve your support. There is a false idea around right now that war will remove a lot of hungry mouths and leave more food, water and wealth for your family. If you killed ten million people that would only leave six billion nine hundred and ninety million left alive. War is not a solution for any of your problems.

There are ways to avoid war, and we recommend that everyone learns to practice these in their lifetimes. Get to know people everywhere, as even on the far side of the globe they are the same as you. They take pleasure in the same things you do, like a child's first step, and are sad at the death of loved ones. When you separate the personal from the political you will not want to fight these people. What do you or they have to gain? Perhaps it is your own government who will benefit from war. Because it is closer to home you can have more effect by direct protest. During the Vietnam war draftees left the country rather than fight. There are ways around fighting, but they depend on resisting the new wars. An entire generation refusing to fight will cause problems to those who want to raise an army.

We have mentioned that those born after 1986 are carrying a lot of light in their persons. Not all of them, but it was the plan that there would be enough to create change once they arrived. These young people are going to be harder to fool as they grow older, the fact that they carry a bright light will begin a process of uncovering hidden secrets. They are also the generation that has nothing to lose, the first generation of young people to be discarded by their governments. Why did that happen? They will be free to act without fear of losing their jobs because they are the unemployed, and the rewards of being inside the institutions will not be theirs. They are outsiders, and if they have a job, a large number of their friends won't. It's very risky to create this strong group of outsiders, but when your eyes are fixed on money you don't always see people. These young people have no stake in prolonging the present society where a few benefit and the rest pay for it. Their non-involvement with the established society will pull the supports from under it. They will not be part of the establishment, and it won't be realised until too late that without them it can't continue in it present form

Here is your ready-made force for change, and in the next few years this generation of young people will see when they are being lied to. They will either turn and walk away from the system altogether, or put themselves in a position to tear it down from the inside. When we talked about the natural dying out of older generations, one day this will be the generation living in a world of light they helped to create. It will be built on different foundations from the old world. Excluded and exposed to the gross inequalities of the current systems they will not hesitate to completely change them. As people of light they will not put into place more systems designed to divide humanity, but choose to knock down the barriers. When the overall human soul chose to make a difference and get the game back on track it incarnated an army of light workers as part of the plan. The human soul group could see the new Earth and the new opportunities she would bring and didn't want to be stuck in the same old mud.

Your young people are here to rescue you all.

Part Five
Ascension

21

The Story So Far

WE WOULD LIKE to summarise the points made in the earlier sections of the book. The Earth is a living planet and a member of the planetary soul group. Planets are beings of light, and they host a number of games that help other souls understand their purpose in life. The Creator learns about Himself by dividing into a multitude of lives across the Universe. You came here to have this experience and designed a game where you were blind to the higher dimensions, to make it harder for yourselves and explore the strength of your soul even when handicapped in this way. This game design contained a terrible flaw, because it left you completely vulnerable to attack and interference from dark angels. You were not aware of this and did nothing to resist their influence, and the Earth herself suffered as a consequence. In the beginning the game design was influenced by the dark angels themselves, who played to your pride and love of extremes. You would find God through the most difficult set of circumstances ever!

This game has been progressing at a very slow rate because every time you incarnate you forget that you are all one soul splintered into many different bodies. Once you grasp this fact you are close to personal ascension. Humanity has been very busy killing each other and trying to grab the best of everything for themselves, leaving others to die in poverty and from lack of food and clean water. In the last 26,000 years little of real importance has been learned about who you are, and you have repeatedly focused on accumulating as much wealth as you can.

Ascension

This is the final year in this phase of your game.

As angels of light we teach but do not interfere, and only the Creator existing inside each individual has the right to act inside a game. He/you perceived the need for help for the planet and for humanity, and a plan was put into motion. The difficulty stemmed from the blindfold the dark angels' manipulated you into agreeing to at the very beginning. The Creator timed an intervention to arrive with the end of the 26,000 cycle, and sent a breath of light across the universe to clear this planet completely of angels of darkness. When these are cleaned off you will be left with a planet of light, and the light and darkness that exist in your own hearts. The game will proceed against a background of light, and many of the veils are dissolving that keep some of the higher dimensions hidden from you. There are many people now who can see more than three dimensions in daily life. The animals and plants always could see all the dimensions.

You have experienced thousands of years of warfare and created a foul energy of hatred and violence, and this has knock-on effects on everything living here. The fractured energy has helped to confuse you and keep you from growing as a species; it will also be cleaned from the planet. There is great potential for a fresh start for humanity.

The Earth has been blanketed in darkness and it has prevented her from being connected to her soul group and being supported by the family of planets for millennia. Where once her great stone circle at Avebury was maintained and used for two-way communication, it was neglected and shut down. The re-opening of this circle allows communication with other stars, planets and all life of the universe to recommence, and she is able to broadcast information in return about everything on this planet. She will be supported by her family again and be more stable, giving you greater stability yourselves as you live on her. The information she receives will be about other souls and what they have been learning as a part of the Creator. There will be some technological advances, but the greater knowledge will be about the Universe, and the attributes of the Source of all life.

The end of every 26,000 year cycle brings a fresh beginning to the Earth, and she has almost finished preparing her new body. Elementals have been building a new Earth in the higher dimensions that is the same as the old, and the energy supports the Earth, not human cities or roadways. These cities will replicate but they will be dependant on human energy alone, because they are not in touch with the planet. Cities keep people together, but separated from other life forms. The planet will find herself recharged by the experience as the focus of the energy from the Central Sun and the stars clustered in the centre of the galaxy. As she is stronger, her light will shine more brightly from within, and it will make it harder for secrets to stay hidden with light from above and light from below. It's time for cleaning out dark corners and old secrets.

The Earth will leave the astrological house of Winter and introspection, and move forward into Spring. There will be an explosion in the energy of new growth and fresh starts on the planet, and turmoil as the old is swept away. If it is not love, and not light it will have trouble holding onto its place on the planet. Animals and plants are already feeling the joy of the coming days, and humans will finally begin to play their game as they originally intended without outside interference. You will ascend to light or descend into darkness on your own merits. Humanity is a soul of light, and to bring about a fresh start now you brought onto the planet your most able people, those in whose hearts the light shines the brightest. Many of these are young, born since the year 1986, and are all here together to work as a unit and to resist those who hearts are filled with darkness. These are the ones who will walk away from war and from the practices humanity uses to enslave one another. These are also the ones who have love in their hearts and have not come to destroy anyone, but to resist in a peaceful manner. Your brightest people are here to help you all make a fresh start in 2013.

The years after 2013 will see many human institutions fracture, and some people will spend time and effort trying to prop them up. These same institutions will be remade in a new way if they have a purpose, such as schools and healthcare. They will be set up in an open manner, and there will be transparency in the way they are run. Many superfluous additions will be dropped, and common sense will prevail. Those institutions that exist in a web of dark secrecy will be thrown open to inspection, and many will not survive, as they will have no purpose. The manner of government will alter to serve the people, and as that happens more of you will take a personal interest in it, and there will be wider involvement at an amateur and service level. Sharing the workload between you will help keep governments open, and give you more time to be outside living with the planet. She will help keep you happy and balanced.

More than anything, an increase in light is an increase in joy. You will not stay in an unhappy situation; you will make the changes necessary for personal happiness whether it is location, job or partnership. People will glow with happiness, and that is another way to increase light in your surroundings and on the planet. By working to maintain happiness in personal lives, others will find it easier to be happy. Duty, for instance, will not be considered a good reason to remain unhappy. This will be an enormous adjustment for you all.

After 2012 the human race will live on a planet of light, and be reconnected to the outer universe. Information will be absorbed by all beings living here, including yourselves, about universal life that considers you as part of itself, and supports you as part of the whole. This will result in many new ways of thinking, and new choices being made on every subject imaginable. The old ways of governing and doing business will be rejected, some faster than others. However many years it all takes, at the end you will be ready for the next step.

22

The Underlying Energy of Cities

THE NEAR FUTURE is a time of change and turmoil as you clear away all the things that were not constructed of light. They are shaking now with instability and have been patched and propped up for decades. Some will crash dramatically, and others you will gently pull down and replace with something that works honestly, with common sense and truth at the core. The feeling that someone else is responsible for you and your life will vanish across society. The link between actions and consequences will be re-established, and lessons will be learned again. When that happens you are back on track in the game of life. Hope will be renewed that life will be fairer, and that government will be fair for everyone.

For the first time you will all recognise the same problems and work together to create something better. Some of you have worked very hard at trying to expose the reality of what is happening in your societies, and tried to engage the interest of others. You felt that your words fell on deaf ears, and that all were satisfied to live in the same old ways. It's not that people were content, but they had no hope of establishing anything better for themselves. There is a deep-seated wisdom in the greater human soul that knows the true state of affairs here and realises your contribution to the condition of the planet. It won't take long for people to join together and begin to create something new, as humanity is a very creative soul.

2012 set the stage for change by withdrawing the energy that supported the old structures. Just like on New Year's Eve, the old year's energy fizzles out and fades away. The knowledge that energy is everything, that any physical object or being is assembled around an energetic structure is the key to making sense of your new world. You have corporations and banks, and they will be missing the energy that holds them up. How do you feel at the end of a long tiring day? You feel like sitting down and letting a chair support your weight, and you lack energy. If you had to keep going on your feet for days you could run out of energy and collapse. We are talking about something similar with your old structures. Some businesses and buildings are strong and flowing into the new energy already; they will not disappear but go from strength to strength. Others are going to run out of steam quickly. They will try to grab energy from any source they can find to keep going, but we say now that it would be throwing good energy after bad. You can let them go and focus instead on creating something new that flows with the new energy. This will create new types of businesses and new opportunities for employment. If you want this to arrive more quickly, support these new ventures by buying from them, and leave the others without your money. It changes the energy a little at a time, because money is energy. Money needs to flow.

You have built a number of overlarge cities in your world, and some people benefit from life there, and others are struggling. The best part of cities is that you are very social and you enjoy being in busy places with lots of other people. You feel more in contact with others as you live in close proximity, and hear their voices even if you don't know them personally. It helps you feel connected and less lonely. The worst part of cities is people's lack of contact with the Earth. You are isolated from your caregiver, the one who agreed to help you while you were resident here, and that cuts you off and makes you feel alone. If you were to take a new-born baby and remove it from its mother and let it raise itself to adulthood, what kind of person would you expect it to be? Some of those feral characteristics are common to humanity, and you mistake them for normality. You forgot about your home and your Earth-mother, and lived without rules or responsibilities, and thinking only about yourselves. In case you think some people "live like animals" we would like to point out that animals live according to common sense. They don't destroy their environment, water or food supplies. They live balanced with nature.

People have always had cities; they used to build smaller cities of great beauty with greenways, gardens, and water features. They walked everywhere inside of them, and kept horses and carriages outside the perimeter for hire when travelling. This brought everyone face to face when they left their homes, and they knew who was living around them. Their houses and apartments were much smaller, they spent less time in them, and were outside more with their feet on the ground, mixing with others. They got their news through word of mouth instead of from a TV or radio. They sang and played instruments and participated in music instead of isolating themselves from others with either very loud music or iPods. These cities were built over places where the Earth provided a funnel of energy that helped nourish the life and activity above. You can see these funnels today, but they all look too small for the size of the city above. Without sufficient energy rising from the Earth the cities became the places they are now. There are some cities around the world that have no light beneath them at all. They are not happy places.

Your planet is changing and in an earlier section we said she would not be supporting cities energetically. Right now your vast cities are the centre of road networks and public transport, while commuters and goods flow into them daily. The surrounding areas support the cities in this way, and your counties and states support the cities through taxes. You all are connected to cities whether you live in one or not. They are home to museums and culture, concerts and sports and the hub of large businesses and government buildings. These are missing from the surrounding areas, and if you go further away the farming areas can be quite empty. Talk to the young people from rural areas and there is nothing to do where they live. They want to be where everything is happening, and that's inside your big cities.

There are places where roads have been constructed on the Earth, where man has assumed they have the right to build. Some of these roads are in inappropriate places and interfere with the Earth's own being, and interestingly they are the roads with a great number of accidents. There is a feeling of impermanence and shifting energy on these, like driving your car on the back of a giant living snake. This is how man-made constructions can be that do not take account of the Earth and how some of your cities and roads have been built. The Earth herself does not support these roads. For example, the energetic circle of Stonehenge goes far beyond the physical stones, and a busy accident-prone road has been built straight through this larger energy circle where it doesn't belong.

Have you ever been to a good party, where its great fun until the energy starts to drop and people decide to go home to bed? This is what we are talking about happening. The Earth in December 2012 will reform and she is not going to provide the energy herself for your large cities. Just like a party, people will start to drift away. If you look at the population density patterns on the Earth now there is currently an ever increasing proportion in the cities. If you remove cities what would you expect to see? It's going to be time to make a new way of living together. What would make sense for you to create?

23

Living in the New Towns

WHAT WE SEE happening several generations from now is that you have spread out across the land once more. The mega-cities are gone, and the people live in a variety of smaller cities, towns and villages. They are more like the old market towns where you could sell goods and buy from specialist craftsmen and women. Culture and sport comes to you with travelling entertainments, and these are always enjoyed. People have returned, not to some medieval peasantry (remember we said you had already lived through that experience), but to lives that are much less stressful. Your new towns will be full of your friends and extended families, work-filled days (but not commuting) and social evenings. You will have discovered that large homes and a separate kitchen in each one can be isolating. It will be more fun to let those who love to cook run a family restaurant while you enjoy eating with your friends. Some of you find shopping and cooking food a daily chore, while others do not.

What about a big city like London, it seems impossible that it will cease to exist. These large capital cities could still have a use as the home of a pared-down government and business centre, and employees could live nearby and come in to work daily. Not by travelling on a long, crowded commute, but living in pleasant city neighbourhoods. There are special buildings like The Royal Opera House or a theatre and the big sports venues that could remain, but function differently. The tickets for all the biggest sports and entertainment events have been priced out of the reach of many people now. There will always be pleasure derived from seeing your best singers and sportspeople, and a need for a place to go and see them. Smaller towns will still be able to attract a variety of travelling performances, and sports are meant to be played as well as watched. You will enjoy participating in and watching local sports where you know the players as friends. It involves you in a different way, and part of your cult of celebrity right now is that you want to personally know these good sportsmen, actors and singers (male and female.)

Did you ever live with a group of people when you were young, and there was always someone to talk to when you came in? People are social, and you have built yourselves beautiful cages to live in alone, and they take a lot of upkeep. How much space do you need to be alone in? When you are away on vacation do you need more than a bed, an easy chair and somewhere to eat? Could you live with fewer possessions and rooms? You have to look after your possessions or they lose their value and usefulness. Life is easier when you don't overburden yourselves. Imagine all you have is a bedroom, bathroom, sitting room, and kitchenette for breakfast. You go out and eat as many meals as you like and sit with whoever else is present. There is a lot of talk and laughter over meals, and your children socialise face to face with other children, and leave early to play outside. Later you if walk down to the local public house or bar, men and women go in together while their children drift in and out. When young people want to drink their parents or neighbours are there to supervise. Society is mixed and everyone knows each other. If someone is lonely because of bereavement, when they step outside there are friendly faces and conversation to help them through the days, and people to eat with in the evenings. Your towns and villages have become empty now in the daytimes as men and women

Ascension

work outside of the home all day.

There are all different types of terrain in the world, and your towns and farms will take different shapes and forms. One of the newer types will be the tree-top towns, where smaller homes will be attached to the trunks and limbs of trees. There will be levels from near the bottom to the very tops of tall trees, and some days you will not put your feet on the ground. What do you think is down there? Instead of paved-over streets and parking lots, there will be grass and plants. This is how the Earth will breathe, and where some of the food is grown. Flyers will be tethered nearer the top to use when you go visiting other villages. Yours will be the home of the forest crops, the berries and nuts that will be in demand from the plains dwellers, and you will also be craftsmen of wood. Others will provide you with the food that needs full sunlight, the fast runners for sports, and livestock. They will be the adept leather workers or fishermen.

The same patterns of villages will be repeated over and over again, so that food doesn't have to be taken around the world. The nearest rivers and seas will be your source of fish and seaweed, not another continent. You sound isolated but you will not feel that way, you will be finding your feet and learning again what it is to be alive and happy. When you have learned how to live with the Earth in balance, and with respect for each other you will have a base to build upon further. When you went off track a long time ago and began pursuing individual wealth, you lost many things. One of the most poignant was joy, and relief from stress. This life will replace working and commuting long hours, watching strangers on TV or reading gossip about them, to really knowing your friends and neighbours and appreciating them.

Humanity is not the only species here to learn, everything living here has a goal. The Earth collected a similar group for this particular time period, and you work together symbiotically. Other species never forgot their Earth mother, and they were able to pool their soul's experiences. Some of them are finished and waiting for you, and others are still working towards their goal. Outside your homes and businesses there is a great deal of excitement about what is coming next.

24

The Years Before Ascension

WE HAVE TAKEN you to the point where the surface of the Earth is more evenly covered in people, with smaller and medium sized towns instead of the mega-cities of today. That way of living will continue almost until the end of your game here. The object will be happiness for all, and care for all. You will have created societies where you can see the inner beauty of your neighbours and love them for who they are. Only by spending time with other people can you see the light in their eyes that comes from their hearts, and recognise that you are all one. We want you to have this experience and feel the amount of love that exists when many hearts are open and sharing love. You have a real treat in store.

We see families today where no one has time to really know each other. You love your families, yet spend more time with people who don't love you, who are work colleagues or teachers. Humans learn to open their hearts by being with people who love them, and sharing their thoughts and experiences. The family is the safe place to be open and show your true selves. You have been robbed of something really precious by working long hours outside the home for more money to buy ever bigger houses, cars and holidays. You've been robbed of the person you could have been, living the contented life you might have had. Children would benefit by shorter school days and more time spent learning by playing outside. This isn't about something that has been done to you by governments or rich people; it's something you have done to yourself. Many families do not provide a safe place to remember how to love, and they have become little more than an economic unit. We see the lack of light in families and we are mystified as to how you chose money and goods over love.

One of the largest benefits of the new ways of living together will be that you are surrounded by people you love; your parents, siblings and children. It may sound tribal to you, but tribes are based on an understanding of us and them. Tribes protect "us" from "them" and in the past have been instrumental in keeping people separate. What we want to emphasize is that the conditions that created tribalism in the past will be gone. You will be able to see that strangers are no different from yourselves. Instead of relying on your tribe for protection you can rely on your own vision and what you read in people's hearts. Love will be the background energy, not fear and danger.

Tribalism generated millennia of aggressive fighting to control the most resources on the planet. You may think of the simpler tribes who have the best farmland or are the closest to water supplies, but your largest tribes now are religious. The rules you use to reach God are one of the key ways to differentiate "us" from "them" in the modern world. You have travelled and mixed your gene pools widely in the last few hundred years, and now you can rely on how you practise religion to keep you apart.

There are also tribes in business where men look after each other. They have different origins but most are based on how rich the father is and which school they could afford for their sons. Wealth is contained in these small circles and is passed from one to another and they are very difficult to enter. This is your ruling elite. Outside of the elite others have developed their own ways of rising to the top, and the gang leader or head of a terrorist or criminal organisation have loyal followers.

These are the organisations people fear.

People are afraid now for their safety and there are many security guards and devices to keep you all "safe." You live in fear of attack by terrorists, by thieves and muggers, and by people intent on harming your children. This is exactly the kind of energy that supplies food to the dark angels. In the future we see fear diminishing and these methods of keeping you safe will seem pointless. It may take a few years but you will dismantle scanners and security cameras, or they will simply be pushed into corners and forgotten. Other tribes will no longer feel like they are different from you, or that they need to fight you for control. You will look at each other differently and see the human within.

You already have the ability to read the energy of another person more accurately than any machine ever could. You can sense the warmth of love or the coldness of a threatening heart. When you realise that you can read the truth as well as any scanner or lie detector you will lose your fear of the unknown. The truly threatening people are very few and they will feel of low vibrational energy to you, and your light will clash with how they feel. You have been plunged into societies of fear by those who allowed themselves to be used by dark angels. They saw they could increase their own control and make more money from fear as you were manipulated and scared stiff. You will not miss fear when it's gone.

The biggest change in creating fear will be the removal of the dark angels from this planet, and the layers of fog and mist they use to surround themselves. You could not see them, and you couldn't always tell how their presence felt or where they were located. They will be blown away on the wave of light and you will see more clearly the good and bad energy in individuals. Once you can see a person's energy it will be like reading their intentions through their hearts. If they are selfish and conniving, it will emanate from them and it won't have the warmth of love. They can smile with their mouths, but you will be able to look and see past the mask and read them accurately. At that point we recommend walking away and associating with those who carry love in their hearts. It is important not to support them with your energy and money, or become one of their followers. They do less damage if no one pays any attention to them.

This ability to read another human's heart has always been possible for you, but you have been laughed in the past at when you say "He doesn't feel right." They want you to prove and have evidence of what's wrong before you act. Here we are saying that your feelings were the most accurate measure, and that proving something by the way you feel is difficult. Therefore just walk away and let them work alone, you don't have to take any other action than that. You will not be acting to harm them, destroying their reputation, or copying their dark behaviour, you will be true to the light in your own heart through the actions you take. We care more about guarding the light in your own hearts than dealing with those who look to harm people. If you act with love you will have more impact than in any other way, and you will keep your own heart safe.

Humanity will live spread out across the towns and villages of the Earth for hundreds of years learning what it is to be one with each other, and one with the Creator. It will become easier to live happily after the days of change are over, and you have settled into the new patterns. There will be increased movement of people from place to place as ease and flow are more accepted into your lives. People will rediscover their telepathic abilities and feel closer to loved ones who have moved away. Crystals will warm the buildings again with thermal energy, just like in Atlantis. The towns and villages will remind you of some of the animal societies like a flock of birds or a school of fish as

Ascension

people connect on many levels. They will feel less alone as they begin to work alongside others intuitively and more efficiently. They will then have more time for their personal hobbies and interests. No one will be bored or lonely, or forced to do a tedious job all day long.

Perhaps you think this sounds too simple and primitive, isn't that how monkeys live? We are not looking at how you live; we are looking at the levels of light and dark that you create. All your hard work, all your expensive items, do not create very much light and happiness. Young children give off the most light, after that we see ever deeper shades of grey until you die. You have not created lives that focus on being happy. If you think there are some happy people, we agree and we can see them by their light; but they are far less than one percent. Your chances of being happy living by today's rules are pretty small, and most of you support the one person on top of the pyramid that is happy. You swapped happiness for believing in the possibility of being happy if only you could – what? Have more money, a partner, a better job, or a bigger place to live? Happiness is light and light is found in your hearts, not in anything that comes from the outside.

To access the love in your hearts has not been easy on the old Earth. Your tender hearts were vulnerable to being hurt, and many of you learned to keep your heart hidden and safe from pain. There will be an adjustment period, and you will begin rebalancing. Those who feed on pain will have gone, and you will be left with just yourselves. Many people who say hurtful things right now feel remorse immediately, they are not bad people. They are under pressure from trying to live in almost impossible circumstances. The adjustment period will give you relief from living such hard lives, and help you balance. By balance we mean a variety of work through the day, people you love near you, stress gone, and laughter reborn.

Ascension

Those of you who are here right now, the most able people that humanity could assemble in physical bodies, are strong. You have the strength to not only live through a period of turmoil; you have the strength to keep your eyes on the light. You will be challenged by events, but as you see the end of institutions built on darkness you will say "Good riddance, it was time for that to go." You will be full of ideas about what you want to see put in their place, which will often be smaller and simpler versions. You will be the first ones to learn that you have more in common with strangers than differences, and find the "us" and "them" differentiation pointless. You fought to be here at this time because this is the most fun an incarnate soul can have, to create a new planet. Side effects may include learning enough to ascend as an individual, or flowing from one exciting time to another and never being bored, and always landing on your feet. You may have felt your circle of true friends was a bit small and others didn't understand why you were interested in the planet, but that's all about to change.

Angels of dark and light belong in your universe and we have roles to play given to us by the Creator of us all. Dark angels never play by any rules, and that's been a problem for you on the only planet where you couldn't see them or recognise their influence. That part of your game was ill conceived, and allowed them a free hand here.

25

Ascension for All

IN THE FUTURE human beings will have reached the understanding that they are all one soul in many different bodies. What has been so hard to learn for so many, many years will have become obvious, and rather than look back unhappily at past mistakes they will be looking forward. They will break the barrier that prevents them from using their group mind and pooling their experiences. Each life will be separate, but with access to all the knowledge and memories of everyone's previous lives. Humans will have reached the starting point for every other species' game across the whole universe and will progress very quickly after that. You will have had a lot of living experience by then, and as a combined soul your game here will be very close to the finish. Your game was to realise that your own soul of light exists as a fragment of God, but whole and complete as all parts of God are. If He were a beach made of sand, and you were one grain of sand, that is your relationship to your Creator. If you are a grain of sand, then all the lives on your planet are the other grains of sand in your little area, including the Earth. When you grasp that you are one with the life on your planet and you are all ready to ascend, the Earth will take you with her.

What does ascension mean exactly? You have lived through experiences here that seem to have taken forever in Earth years. Before that you were on a different planet, and even earlier on another planet. You found your relationship to God already, more than once. What you were trying to achieve here was to see if you could do it blindfolded, and that's ok; the universe exists just to have experiences. As we said, the most extreme game ever! You made progress every time you arranged a game, and you were considered a safe pair of hands. If you couldn't find God wearing a blindfold, then no one could. The Earth agreed, but she could see the risks to herself, and we could see them also. She didn't have many ways to save herself if everything went wrong. That is why she asked such a high price, that in payment for this experience you would join with her in ascending as one large soul. You agreed, and you agreed to end your experiences by ascending. After all, what could be more challenging than the blindfold game?

The planet was filled with many, many tiny species that were ready to join her in ascension, and they found your game troublesome, to put it mildly. You were so unknowing that you killed many of these by hunting them to extinction. That is very rare in the rest of the universe, but it has happened a lot here. You tore apart the physical surface of the planet and made many species homeless and they starved to death. Antibiotics and pesticides wiped out many more species. They watched you multiply until there was little space or food left for them. Nevertheless, they had signed up to a game leading to mass ascension. When we wrote about the breath of God being sent to clear this planet of dark angels, you need to consider it from the aspect of the many souls suffering here because of one soul's game. The Creator was willing to help in a game that was neither fair nor just. You are being helped because you are loved, and because the Earth asked for help. Everyone wanted to see how this game would turn out, and there is no one blaming you or thinking you are bad. Being blind to the higher dimensions for the entire human race was unfair. It was like you were all physically blind and there were others preying on you for food. How much chance would you have? And yet, this is an accurate description of what has been happening here. When the hunters are gone and you have a chance to live without them you will create positive change.

When every soul group here is ready to ascend the Earth will drop her physical body while at the same time you will all let go of your own. You will have finished, and return to your energy state. All remaining species here will be in their energy states of light at the same time with the intention to combine. You will let go of your separateness and blend, and share everything you have learned over all the lives on all the planets. There will be a very large soul of light, intensely bright and wise, and knowing exactly who they are. Energy is not limited to one spot in a galaxy, but can be anywhere in the universe. Wherever it goes it increases the level of overall light inside this contained space. And what does light do? It shines into dark corners, and it shines into the spaces between the galaxies. It dissolves the dark with love.

Angels of light influence planets and areas of space, and we have been helped by the stars and beings of light. The dark angels and beings of darkness also influence planets and the dark spaces farthest from the stars, the deep space between the galaxies. Over time what was once fluid and moving has become rigid, and a planet seldom changes from light to dark, or dark to light. The risks involved in hosting a difficult game began to be seen as too great for a planet, and the universe has been heading towards stagnation. Humanity initiated a risky game, and the Earth responded with a request to join her in ascension. The day of the Earth's ascension is the beginning of the universe pulling back towards the Creator. The new combined soul of light will change the balance in this universe, and will bring it closer to the end.

The End

The End

IN THE BEGINNING you were a thought and an idea of the Creator. He existed alone in timeless space and, having been created by His Creator, He did not fully know Himself. Alone in darkness He existed, and any others of His soul group were isolated from Him. He was playing His own game of self knowledge through isolation. What did He do?

As a beach is a single place made of many grains of sand, our Creator dispersed into many universes and learning experiences. You ended up in your present universe, but how did you get here? He had a thought of universes, and then a thought of how He wanted to learn about Himself. He created angels and asked them to choose between light and dark or between happiness and pain, and sent them as team leaders into each universe. Angels are vast beings of light and do not need to incarnate, as it is not their purpose. Dark angels are vast beings of light who faithfully bring to God the lessons He learns from pain and misery. In this way we do His will as He requested of us.

In ourselves we angels know that we are all part of God, and that we are equals in every way. It is our purpose to oppose each other and to bring to the Creator lessons learned from the heights of happiness or the depths of despair. If we do not encourage you to reach these extremes you may never reach them. So the dark angels settle on a planet and subvert every good thing we, the angels of light, have tried to put into place. Where we think there are rules of behaviour, the dark

The End

angels are around the corner ripping up the rule book. They are very ingenious, but then so are we. We sound like we could have played this game alone over eternity, but we were sent here to marshal our teams and play a very large, universe-wide game. You and many other souls of light, grey (neutral) and black inhabit this universe to learn by living or watching and influencing others. There are many different roles and they do not all involve incarnation on a planet.

There were many and varied souls who chose to go onto planets and live, and bring their tears and laughter to God. You are one of these many souls, and every soul is different. Each soul has grown in knowledge as they finished a game and moved on to another planet or to simply exist as an energy body. Many of the souls present on Earth right now have had lives as different beings. You humans have lived on planet after planet, and you are strong and forward looking. You have had troubles before, and always you overcame them. Humanity is one of the most creative souls in the universe, and you put amazing twists on everything you choose to do.

Your creative soul helped design this difficult game, and it took a strong soul to feel they would be able to finish it at all, much less finish it in light. It's been a risky and lonely game. Today you are repeating a pattern that has happened over and over since you began here; you are moving apart economically to the haves and the have-nots. This has been your method of exploring the edges of happiness and pain, and all the space in between. You have gifted this back to your Creator to learn about how far a game can go in polarity, or in two directions at once. You balance light and dark on a knife edge and all beings here learned with you. There is vibrancy and risk in every minute of every day on Earth. You designed a game to learn everything that could be learned from joy and pain, and to do that the blindfold was necessary.

The End

We said in *Planet Earth Today* that the game played here was taken to such lengths that it would never have to be repeated again. It took a long time for the residents in this Universe to come up with a game that would put a full stop to the learning experiences, but humanity did. You allowed pain to enter with the blindfold, and you hoped that your own hearts of light would get you through.

Earth agreed to assist you, the only planet that accepted in her early days another consciousness as part of her being. The dual-soul planet of Earth and crystals is stronger and shinier than any other, and would be the planet to help you and play host to your daring game. She was happy to begin with, until the open door to the dark angels caused her too much damage. She was not free to throw them off and ended up being enslaved herself. It has not been a happy experience for her, and she would avoid allowing them residence again. But she has help now in the form of the greater universe, and is free of her tormentors long enough to give you a breathing space and help you create some joy of your own. You have entered a new phase together.

There is no wrong ending to this game for you as all experience is to be handed back to God. If one day you rejoin the other life here as one being and you will know more about what it is to be alive and live life to the absolute maximum. As we said in this book your life is about your emotions and the light that shines from your hearts. Do you radiate love and light, or fear and sadness? Only this matters.

There are planets in this universe who hosted games ending a long time ago in happiness and joy, and sealed themselves off. These happy planets have remained as planets of light for a long time, unchanged, with happy souls living on them. They have anchored light and helped keep the universe balanced. To visit one of these planets is a joy, but there is little for us

The End

to do there; for souls who watch and never incarnate there is pleasure but it's no longer very instructive. Similarly there are also the opposite planets of warfare and slavery. Earth is the planet where everything is constantly changing, if you looked away for one hundred years of your time, and turned back you would hardly recognise people's lives in the material sense. Their energy has fluctuated from light to dark over that time, with a steep dive in more recent years.

There is a reason why a planet would seal itself off and refuse to host more games, and that is fear. A planet can be killed, and that or the fear of a long unhappy game has caused them to withdraw. Opportunities to continue learning were greatly reduced, and the universe had lost its vibrancy. The Earth accepted a risky challenge, and demonstrates living to the full extent of her being.

We do not watch you like a popular television program, we observe you carefully like a doctor with a patient in intensive care. As a doctor, we have to rely on the strength of the patient for a good recovery. We watched as economic differences increased and more and more people were killed, often because they were in the wrong place at the wrong time. If they were the wrong tribe in modern Sudan, or the wrong religion in Germany in World War II they were eliminated. After hundreds of thousands of years, the unhappiness and darkness seemed to be winning. It picked up speed and the rate of warfare and death of civilians increased.

We wrote that if you stood by and watched a government kill its people and did not help you were complicit in the deaths of those people. We were in the position of watching a soul of light, one of us, cannibalise itself on a planet. We have dark brethren and we were letting them ride you like horses over a cliff and we stood by watching. This game was to extend the distance between light and dark and explore the space in

The End

between, as well as the extremes. If we did nothing the game would collapse before you had a chance to play to the end.

Is it fair to hijack human bodies at every opportunity and make you act on instructions? Whose game is it then? This had turned into a new game of dark angels versus blind humans and the rest of the universe felt there was nothing more to learn from this. Other planets' residents had not been blind to the light and dark angels, and when a demon approached they could see it clearly and know who they were dealing with. They were able to choose to work with light or dark with open eyes. (It's a little like your stories of the devil tempting people into trading their soul for talent or wealth.) The entire universe used its own powers of creativity and arranged some help. The object was to put you back on an even footing where you are guided but not controlled unless you choose to be. You continue to have free choice.

In 2012 the planet will be blown free of controlling demons by light in many forms. The first time was in April, the last in December. There are men and women who will choose to act as they always have. Others will find it far easier to act in kindness and love and bring those attributes forward into the world. When you tried in the past your efforts were often subverted. You will be able to be who you choose to be for the first time since all the trouble started in Atlantis. Life will once again be what human beings make of it.

You will be put back on a level playing field and begin to see many changes immediately. The first will be that you have more perception and clearly see truth from lies, as the fog around the dark angels will vanish when they go. You have areas on this planet that are always fighting and that is because they are infested with these beings, and they live on aggression and pain. They actively fuel these emotions in people. It could be whole countries or just a rough neighbourhood pub or bar.

The End

People will have behaviour patterns, but the push to express them will subside. Those dark angels who sit on the roofs of your seats of government will be gone, and if you can change the way they do business they won't have the drive to revert into lies. You will be able to elect representatives based on who they really are, without the mask or disguise they've put in place. No one can mask their heart.

It's been a very long game here for the Earth, and for humanity. You showed how far a game could go, and blindness to the higher dimensions was a step too far. In spite of that drawback you balanced great pain with great joy in individual lives and never gave up looking for love. The heart-breaking loneliness of incarnating separately will not stop you realising that you are all one, and quite honestly, that is some achievement. We are immensely proud of you.

We angels of light will be available to guide you as we have always been, and the dark angels will also be here to consult. They will not hijack any more bodies; that will be finished, and they will only enter when invited. Both sides will guide through teaching and helping when asked. You are a soul of light, and although you will hear of some people dealing with darkness, by far the greater number will never do so. Your greatest challenge in the next few generations will be to align the actions of the world with your soul of light. Working together as one is how your game will come to an end. When you ascend with the Earth the universe begins its return to the Creator. We have written this book to encourage you to move ahead with joy in your hearts as you live on this beautiful planet.

Books by Candace Caddick

In 2009 the Archangels wanted to write a channelled book about the Earth, and help us to see the reality of the world we live on. *Planet Earth Today* shows a sentient planet of incredible beauty, and a human soul of light that is under attack, lied to and deceived. I channelled this book by six Archangels, which was a combination of them explaining and me asking questions. We collaborated, and I would say that seven of us wrote it together. *Planet Earth Today* is the first book of a trilogy including *The Downfall of Atlantis* and *And I Saw a New Earth* that the Archangelic Collective and I have written. The contents of their books are always relevant to what is happening now as the Archangels have so much to teach us.

There is a single story of humanity, a golden book like a long scroll and the three books have been taken from here and typed up. I felt that as long as I was learning new information when writing, information that I couldn't begin to make up, I was on track as an accurate channel. I watched the flow of golden words enter the computer each time until it was the last page of the book. After that my daughter and I checked and checked that I had written it correctly, each paragraph and line examined to see if the golden energy ran through it steadily or if it wavered indicating that it was not quite accurate. Only when we were happy was a section considered complete. Later sometimes I would add more clarity to a section, as my own understanding improved and I could put in more detail. I channel using a combination of sound and sight, whichever is quicker, and where it is written the best I have been writing down their words.

Planet Earth Today

The first book gives background information on the roles of Earth and the human soul in the universe. Life is experienced so that they can know themselves and learn why they are alive. Humanity wished to live on Earth wearing a blindfold; they could see neither the higher dimensions or connect to their greater human soul. This has led to great loneliness and separation as you began to play the hardest game ever conceived. The Archangel of Darkness presents his point of view of humanity on Earth, and the Archangels of Light: Ariel, Esmariel, and Hophriel write with techniques to take you forward with hope.

This book serves as the introduction to the trilogy as it takes place before the other two books in time, and the information there about the planet or Atlantis is not repeated in any other book. However, each book stands alone and can be read individually.

ISBN 978-0-9565009-0-8

The Downfall of Atlantis

In the story of humanity on Earth, the time spent living and learning on Atlantis cannot be ignored. During those long years the darkness gathered around human beings, and science developed a heartless approach. There were slaves made of combinations of animals and people and ultimately cloning to keep the wealthy and important alive indefinitely. Cloning was the final crack in the system that led to ruin and the end of Atlantis.

Those who refused to go along with the new science escaped the end and settled on the surrounding land masses forming the new post-Atlantean civilisations. The Atlantean influence is explored in the cultures of Africa, Egypt, Britain and Celtic Europe, North, Central and South America. They learned much from these people in return.

Their civilisation remained intact for a long time in Britain because of the ancient sites of power at Avebury, Stonehenge and Glastonbury Tor. When the Shadow in the East pushed westwards into Europe the light of these venerable societies vanished until only the now mythical King Arthur and Merlin were left to protect the Earth from darkness. Their story explains the true significance of the great stone circles, and how we came to forget the real story of Arthur and the sacrifices he made to destroy the invading armies. The connection in a straight line between Atlantis, post-Atlantean civilisations, King Arthur and the Time of Legends is explored so we can remember those things we have forgotten, and not repeat past mistakes.

ISBN 978-0-9565009-1-5

Books available from online retailers.

About the Author

I am a teaching Reiki Master who studied for ten years with my Master before being initiated in the Usui Shiki Ryoho system of Reiki. During the course of the twenty years I've been practicing my own Reiki, my ability to channel became clearer and stronger until a few years ago I realised I was able to see the world around me in a way that others did not. My efforts as I worked with my own archangelic guides as a channel was always to unblock and be clear, with no preconceptions of what they may say next; to stand well back and just watch and listen. My guides introduced me to certain people they wanted me to help, where all questions could be answered immediately through channelling, and these people would be ready to play their roles in 2012. These people would then move on and begin their work for the Earth.

More information and a regular channelled blog can be found at: www.candacecaddick.com.

www.ingramcontent.com/pod-product-compliance
Lightning Source LLC
Chambersburg PA
CBHW032042090426
42744CB00004B/99